Histor

A Place in History:
A Guide to using GIS in Historical Research

Ian N. Gregory

Oxbow Books 2003

Published by Oxbow Books for the Arts and Humanities Data Service

ISBN 1 84217 036 8
ISSN 1463 - 5194

A CIP record of this book is available from the British Library

© Ian N. Gregory and Oxbow Books 2003

Cover illustration © Thanks to the staff and students at the Department of Geography, University of Portsmouth who contributed to the images shown on the cover.

The right of Ian N. Gregory to be identified as the Author
of this Work has been asserted by him in accordance with
the Copyright, Designs and Patents Act 1988.

All material supplied via the Arts and Humanities Data Service is protected
by copyright, and duplication or sale of all or parts of any of it is not permitted, except
otherwise permitted under the Copyright, Designs and Patents Act 1988. Permission for
any other use must be obtained from the Arts and Humanities Data Service.

This book is available direct from

Oxbow Books, Park End Place, Oxford OX1 1HN
(Phone: 01865-241249; Fax: 01865-794449)

and

The David Brown Book Company
PO Box 511, Oakville, CT 06779, USA
(Phone: 860-945-9329; Fax: 860-945-9468)

or from the website
www.oxbowbooks.com

Printed in Great Britain by
Information Press, Eynsham, Oxford

Contents

EXECUTIVE SUMMARY vi

ACKNOWLEDGEMENTS vii

SECTION 1: GIS AND ITS USES IN HISTORICAL RESEARCH 1
1.1 Introduction 1
1.2 The terminology of GIS 1
1.3 Uses of GIS 3
1.4 Problems with GIS 4
1.5 The GIS learning curve 5
1.6 Towards good practice in GIS 6

SECTION 2: THE WORLD AS VIEWED THROUGH A GIS 8
2.1 Introduction 8
2.2 Attribute data 9
2.3 Vector systems 11
2.4 Raster systems 13
2.5 Other systems: terrain modelling with TINs 14
2.6 Bringing it all together with layers 15
2.7 Conclusions 16

SECTION 3: ACQUIRING SPATIAL DATA 18
3.1 Introduction 18
3.2 Scanning maps to produce raster data 19
3.3 Digitising maps to produce vector data 19
3.4 Geo-referencing 20
3.5 Error and accuracy 21
3.6 Digitising attribute data 22
3.7 Raster-to-vector and vector-to-raster data conversion 22
3.8 Primary data sources 22
3.9 Buying data or acquiring it free 23
3.10 Conclusions 24

SECTION 4: BASIC GIS FUNCTIONALITY: QUERYING, INTEGRATING AND MANIPULATING SPATIAL DATA 25

4.1 Introduction 25
4.2 Querying data 25
4.3 Manipulating and measuring spatial data 26
4.4 Buffering, Thiessen polygons and dissolving 26
4.5 Bringing data together to acquire knowledge 28
4.6 Formally integrating data through overlay 28
4.7 Integrating incompatible polygon data through areal interpolation 33
4.8 Conclusions: information from spatially detailed, integrated databases 35

SECTION 5: TIME IN HISTORICAL GIS 36

5.1 Introduction 36
5.2 The need for understanding through space and time 36
5.3 Time in GIS 37
5.4 Methods of handling time in historical GIS 37
5.5 Conclusions 41

SECTION 6: VISUALISATION FROM GIS 42

6.1 Introduction 42
6.2 Mapping and cartography in historical research 42
6.3 Developing understanding from basic mapping through GIS 43
6.4 Producing atlases from GIS 44
6.5 Electronic visualisation from GIS 45
6.6 Other forms of mapping 45
6.7 Moving and interactive imagery 46
6.8 Conclusions 47

SECTION 7: SPATIAL ANALYSIS OF STATISTICAL DATA IN GIS 49

7.1 Introduction 49
7.2 What makes spatially referenced data special? 50
7.3 Spatial analysis techniques 51
7.4 Spatial analysis in historical GIS 53
7.5 Conclusions 54

SECTION 8: QUALITATIVE DATA IN GIS 55

8.1 Introduction 55
8.2 Types of qualitative data in GIS 55
8.3 Case studies 56
8.4 Conclusions 57

SECTION 9: PRESERVATION, DOCUMENTATION AND THE ROLE OF THE HISTORY DATA SERVICE 59

9.1 Introduction 59
9.2 Obtaining data from the History Data Service 60
9.3 Depositing data with the History Data Service 60
    Ensuring preservation 60
    Providing access 61
    Professional recognition 61
9.4 Documenting a GIS dataset 61
9.5 Further information 63

SECTION 10: GLOSSARY AND BIBLIOGRAPHY 64

10.1 Glossary 64
10.2 Bibliography 72

Executive Summary

This guide is intended for historians who want to use Geographical Information Systems (GIS). It describes how to create GIS databases and how to use GIS to perform historical research. Its aims are to:

- Define GIS and outline how it can be used in historical research
- Evaluate the way GIS models the world
- Describe how to get data into a GIS
- Demonstrate the basic operations that GIS offers to explore a database
- Review how time is handled in GIS
- Explain how GIS can be used for simple mapping and more advanced forms of visualisation
- Discuss quantitative data analysis within GIS
- Illustrate the use of GIS for qualitative analysis
- Summarise documenting and preserving GIS datasets.

The book provides a broad sweep of GIS knowledge relevant to historians without assuming prior knowledge. It includes case studies from a variety of historical projects that have used GIS and an extensive reading list of GIS texts relevant to historians.

It has been commissioned by the History Data Service as part of the Arts and Humanities Data Service publication series *Guides to Good Practice in the Creation and Use of Digital Resources*. The series aims to provide guidance about applying recognised good practice and standards to the creation and use of digital resources in the arts and humanities.

AUTHOR DETAILS

Dr. Ian N. Gregory
Department of Geography
University of Portsmouth
Lion Terrace, Portsmouth
PO1 3HE
UK
Ian.Gregory@port.ac.uk

Acknowledgements

I would like to express my sincere thanks to the following people who reviewed this guide, offered suggestions for its improvement, and contributed to its shape.

- Alasdair Crockett, University of Essex
- Graham Jones, University of Leicester
- Roger Kain, University of Exeter
- Karen Kemp, University of Redlands
- Anne K. Knowles, Middlebury College
- Mark Merry, History Data Service
- Donald Morse, University of Edinburgh
- Alastair Pearson, University of Portsmouth

Section 1: GIS and its uses in Historical Research

1.1 INTRODUCTION

This guide is written for historians who want to use Geographical Information System (GIS) in their work. It covers both creating GIS databases, and exploiting the information held within them. A minimum of jargon has been used throughout, and no prior knowledge of GIS has been assumed. It should be noted though that GIS is a technical subject, and some knowledge of computing will make the guide easier to understand. In a short text such as this it is impossible to give a comprehensive treatment to most of the themes introduced. Instead, the aim is to highlight the main themes and provide references that allow the reader to follow them up in more detail. There is an extensive literature on GIS but the literature on using GIS in historical research is currently limited. Anne K. Knowles's book (Knowles 2002) marks the first edited collection of case studies on historical applications of GIS and is highly recommended to anyone with an interest in the field. Beyond this, the literature on using GIS for historical applications is widely scattered. This guide attempts to bring this literature together to illustrate how historians have used GIS. To this end each Section provides summaries of case studies and details further reading on historical examples. As this literature is limited, some studies are quoted in more than one Section but no study is used as a key reference more than once. The aim is to give the reader an understanding of both the structure of GIS, namely the way that it models the world, and an idea of the mentality that should be followed when using GIS for historical research.

1.2 THE TERMINOLOGY OF GIS

An examination of the basic GIS texts will give many different definitions for GIS. The reason for this is that there are two basic ways of approaching GIS. It may be regarded from a tools-orientated point of view that explores how the software models the world, or from an approach-orientated point of view that explores what GIS allows us to do.

The tools-orientated approach describes GIS from the point of view of the software, for example, *ArcView* or *MapInfo*. These are often regarded as complex computer mapping programs. This is a misapprehension: GIS software combines computer mapping functionality with a form of *database management system* (DBMS) such as *Dbase, Microsoft Access*, or *Oracle*. Computer mapping systems such as *Adobe Illustrator* or *CorelDraw* are designed to produce high-quality graphical output. They include functionality such as the ability to draw features, move features from one location to another, change shading and line width, and so on. Although some, but by no means all, of this functionality will be found in GIS software, it is more helpful to think of GIS as a spatially referenced database. As such, the data are represented in two ways. Firstly, there are rows of data found structured in the same way as in a conventional database. In GIS terminology this is called *attribute data*. Many GIS software packages allow attribute data to be stored in conventional database management systems such as the ones described above. The special feature of GIS software is that each row of attribute data is also

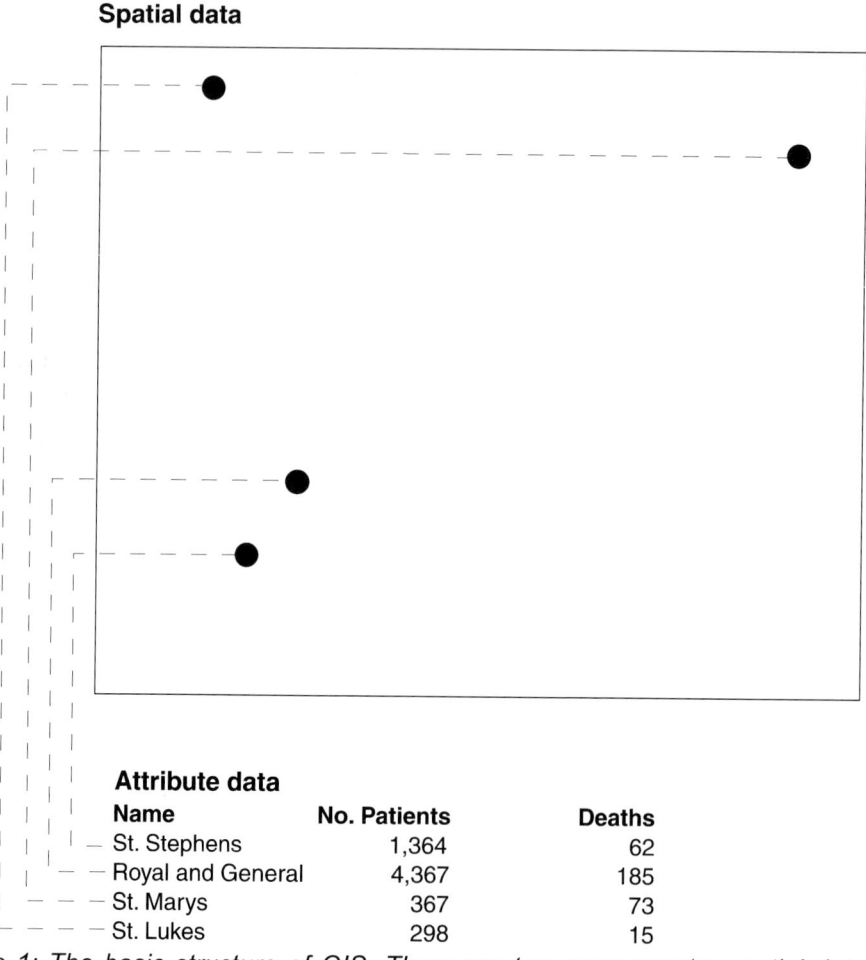

Figure 1: The basic structure of GIS. There are two components: spatial data that show where the feature is and attribute data that provide information about the feature. These are linked by the software. In this example we have some fictional hospital data showing the location of hospitals combined with information on their names, numbers of patients and numbers of deaths.

represented by a spatial feature that is represented by coordinates and is thus mappable. This spatial feature will be a *point*, a *line*, a *polygon* (the technical term for an area or zone) or a *pixel* depending on the type of data it is representing. This is termed the *spatial data*. Using this combination of attribute and spatial data *GIS data* combine information on what an object is with information on where it is located for each feature in the database.

The fact that there are both spatial and attribute data allows the database to be exploited in more ways than a conventional database allows, as GIS provides all the functionality of the DBMS and adds spatial functionality. For example, a user has a conventional database consisting of data on hospitals. The columns in this database include the name of the hospital and the numbers of patients, deaths, and so on. There is one row of data for each hospital: in the GIS this becomes the attribute data. The spatial data are a point location for each hospital stored as a *coordinate pair* but represented on screen using a dot or another form of point symbol. The basic structure of this is shown in Figure 1. This combined representation of the study area opens up a whole new range of possibilities that neither a DBMS nor a computer mapping system could handle on their own: for example, we could draw the locations of all the hospitals, click on one of them, and have the system list its name and other attribute data. We could also select only the hospitals with, for example, over 1,000 patients. In a conventional database all we can do with this information is list the data for the hospitals concerned. In a GIS we can do this, but we can also draw where they are, or perhaps draw all of the hospitals using different shadings to indicate their different sizes. In contrast to a conventional DBMS, therefore, GIS allows users to gain an understanding of the geography of the phenomenon they are studying. Unlike a computer mapping system, GIS provides the underlying data that form the patterns shown on the map.

The above approach demonstrates that GIS software can be regarded as a spatially referenced database. It is able to map the data and also to query it spatially, asking questions such as 'what is at this location?' that a conventional database would be unable to answer. An approach-orientated definition asks how we can make best use of this dual-component data model. This involves considering both the benefits and the drawbacks of including location into our exploration of patterns. In the GIS literature this approach has become known as *Geographical Information Science* (GISc), and Siebert (2000) refers to 'spatial history' in very much the same way. This approach uses GIS as part of the process of exploring change geographically and temporally. 'Historical GIS' is also a term that is becoming increasingly used to describe approaches to historical research involving the use of GIS (Knowles 2000).

One other term that also needs defining in this section is the word *space*. Among the GIS community this term is used in a very similar way to the term *location*. Spatial data are data that refer to locations, and where a GIS book might say 'consider the role of space', a historian may well say 'consider the role of location'. There is, in fact, a slight difference in definition as space is a scientific way of defining location, usually through a coordinate system, thus 'a location in space' usually means a location that can be defined using one or more coordinates.

1.3 USES OF GIS

There are three basic categories that GIS can be used for: as a spatially referenced database; as a visualisation tool; and as an analytic tool. A spatially referenced database allows us to ask questions such as 'what is at this location?', 'where are these features found?', and 'what is near this feature?'. It also allows us to integrate data from a variety of disparate sources. For example to study the dataset on hospitals described above, we might also want to use census data on the population of the areas surrounding each hospital. Census data are published for districts that can be represented in the GIS using polygons as spatial data. As we have the coordinates of the hospitals and the coordinates of the district boundaries we can bring these data together to find out which district each hospital lay in, and then compare the attribute data of the hospitals with the attribute data from the census. We may also want to add other sorts of data to this: for example data on rivers, represented by lines; or wells, represented by points, to give information about water quality. In this way information from many different sources can be brought together and interrelated through the use of location. This ability to integrate is one of the key advantages of GIS.

Once a GIS database has been created, mapping the data it contains is possible almost from the outset. This allows the researcher a completely new ability to explore spatial patterns in the data right from the start of the analysis process. As the maps are on-screen they can be zoomed in on and panned around. Shading schemes and classification methods can be changed, and data added or removed at will. This means that rather than being a product of finished research, the map now becomes an integral part of the research process. New ways of mapping data are also made possible, such as *animations, fly-throughs* of virtual landscapes, and so on. It is also worth noting the visualisation in GIS is not simply about mapping: other forms of output such as graphs and tables are equally valid ways of visualising data from GIS.

Although visualisation may answer some of the questions a researcher has about a dataset, more rigorous investigation is often required. Here again GIS can help. The combined spatial and attribute data model can be used to perform analyses that ask questions such as 'do cases of this disease cluster near each other?' in the case of a single dataset; or 'do cases of this disease cluster around sources of drinking water?' where more than one dataset are brought together. To date, this form of analysis has been well explored using social science approaches to quantitative GIS data. It has not been so well explored using humanities approaches to qualitative data, but this is one area where historians are driving forward the research agenda in GIS.

1.4 PROBLEMS WITH GIS

It is important to note at an early stage that there are also serious limitations to GIS. These fall into four main classes: problems to do with the GIS data model; problems to do with the data themselves; problems with the academic paradigm; and practical problems.

Spatial data consist of one of four types of *graphic primitive*, namely: points; lines; polygons; or pixels. Where the data have precisely defined locations that realistically represent the features to be modelled, GIS is a powerful tool. Other data cannot be adequately represented spatially in this manner. This may be because the data do not fit the four types of graphical primitives well, or because the data are imprecise, a problem that GIS cannot cope with easily. For example, GIS is well suited to modelling hospitals and census districts in the manner described above, but is not well suited to representing the catchment areas for the hospitals where these are poorly defined and overlap heavily with surrounding catchments.

Secondly, the data themselves can also cause problems. Much historical data will be taken from historical maps, which may not be accurate, and the representation of features from these maps in the GIS at best will only be as accurate as the original source. In reality they are likely to be worse, as new errors are added when the data are *captured* (or transcribed, to use the historical term). Many of the clues about the accuracy of the original source will be lost when the data are captured. An obvious example is that if a feature is represented on a map by a crude, hand-drawn, thick line we may question its accuracy. In the GIS it will simply appear as a digital line like any other. Less obvious, but at least as important, is the scale of the source map: a map is only ever accurate within the limitations of its scale. In a GIS, however, we are able to zoom in hard or to integrate data taken from maps with very different scales. This demands more from the data than the original map or maps were designed to accommodate and may lead to inaccuracy, error and misunderstanding. Although historians will be familiar with issues associated with the accuracy of transcriptions, GIS is particularly demanding of the accuracy of its source data, as will be described in section 3.4.

Thirdly, the academic origins of GIS were located within technological advances in the earth sciences. Its role in academic geography has yet to be fully established, and history trails some way behind this. Through the 1990s there was considerable debate in geography about whether GIS offered a cohesive, scientific framework that could re-unite the subject, or whether it was a return to a naively positivist agenda. From the historian's point of view, GIS offers new tools, new techniques and new approaches. These approaches must be used critically and should complement traditional ideas, approaches and concerns.

The final set of limitations on GIS is practical. GIS software is expensive and may be difficult to use. The cost of GIS hardware has fallen over recent years but can still have a high price, and GIS data are often financially expensive to buy and capturing them yourself is costly in time as well as money. People with GIS training are often very employable and thus expensive. As a result, entering into GIS is often more costly than originally anticipated and should be done with care.

1.5 THE GIS LEARNING CURVE

Moving into GIS requires a long and sometimes daunting learning curve. It involves learning new technology, acquiring or building spatially referenced databases, and learning a new approach to investigating the patterns within those databases. Once the need for a GIS has been identified, most GIS projects will go through three distinct stages. The first stage is the resource creation phase. In this stage the hardware, software, staff, and spatially referenced databases are acquired. As is described in Section 3, building spatially referenced databases is a particularly slow and costly operation. These costs should not be underestimated, and may rightly act as a deterrent when the use of GIS is being considered.

The second stage usually involves basic mapping and querying of the spatially referenced database, using the kind of techniques described in Section 4 and the early sections of Section 6. In this stage the GIS adds new options and new functionality to a project, and allows the exploration of spatial patterns and spatial relationships held within its data. This stage can be reached quite quickly once the database has been built, as the techniques used are well developed and central to most GIS software packages.

In the third stage the use of GIS becomes more sophisticated. This involves the use of more complex explorations, analyses and visualisations. The researcher will often devise new techniques and methodologies appropriate to their own specific data and problems. This can involve sophisticated visualisations, as described in the latter sections of Section 6, and the use of more complex quantitative and qualitative analyses that attempt to explore the data through all three of their components: namely attribute, space and time, as is described in Sections 7 and 8. In this stage the GIS is no longer simply a tool but also becomes an explicit part of the research agenda.

From this discussion it should be apparent that running GIS projects is a middle- to long-term process with long lag times before the full rewards of the initial investments are realised. This fundamentally influences who should and should not get involved in using GIS. Healey and Stamp summarise this succinctly when they state "GIS is not suitable for 'one-off' map production or short-term investigations but is ideal for the steady and meticulous development of geographically referenced data resources that can be utilized for a variety of purposes over a period of years. The initial investment will pay handsome dividends in the medium term." (Healey and Stamp 2000, 590).

1.6 TOWARDS GOOD PRACTICE IN GIS

Throughout this guide the issues, dangers and pitfalls of using GIS will be raised and references given as to where more detail can be found. As the use of GIS among historians is in its early stages, and the potential applications of GIS are so broad, it is difficult to be precise about what constitutes good practice. The basics of good practice when using GIS can be summed up in three broad points:

- Always think carefully about the impact of location. The basic question underlying GIS research is 'what impact is the geography having on my data?'. This applies to both genuine geographical issues, such as the links between industrial development and the development of the transport network, or problems with the spatial nature of the data. These can include apparently simple considerations such as the impact of combining two layers taken from different scale sources, as well as much more conceptually complicated considerations, such as the impact of spatial autocorrelation on statistical techniques.

- Always be aware of the data's limitations. This applies in particular to the limitations of the source material, and to limitations connected with spatial issues such as scale and accuracy.

- Avoid unnecessary simplifications when exploring, visualising or analysing data. This means trying to use the data in a form as close as possible to their original with a minimum of aggregation though space, time or attribute. Where aggregation is unavoidable or is present in the source, the impact of the aggregation on the patterns formed must be considered.

These three points are similar to good practice in historical research generally. The key difference is that with GIS the historian has to consider good practice in relation to spatial data as well as to attribute data, and issues associated with space, location and geography present particular challenges to good practice that the historian is unlikely to be familiar with. This guide will raise these issues in more detail and give advice on good practice in dealing with them.

GIS, in combination with other branches of scholarship, has the potential to provide a more integrated understanding of history. There are, however, risks. In particular, GIS data are expensive, GIS expertise is expensive, and the learning curve for people using GIS is steep. This means that researchers wanting to get involved in GIS should do so carefully, and be aware that the initial investments are high and that the rewards may take some time in arriving. On a more positive note, once spatially referenced databases are built they can be significant works of scholarship in their own right and can enhance our understanding of a problem. This means that "what might initially appear to be simply an exercise in automated cartography proves able to supply detailed answers to questions so detailed and time-consuming to obtain by manual methods that scholars have previously tended to avoid asking them" (Healey and Stamp 2000, 584).

Section 2: The World as viewed through a GIS

2.1 INTRODUCTION

This Section gives a detailed description of the data models that GIS software packages use to store their data. Little of what is said here is directly historical; however, it is important to present a brief overview of the way in which GIS models the world. The Section also establishes the terminology used throughout this book, as this varies between different authors and different software packages. There are many good introductions to GIS that are appropriate for historians. In 1991 Maguire, Goodchild and Rhind edited a large collection of papers on GIS in two volumes that rapidly became the standard reference work in GIS (Maguire *et al*. 1991). In 1999 a completely re-written second edition of this book was published, edited by Longley, Goodchild, Maguire and Rhind (Longley *et al*. 1999). Most of the first edition of the 'big book', as it became known, was then made available online. In 2001 the four authors responsible for the second edition produced a monograph on GIS that provides a wide-ranging introduction to GIS at a more affordable price than the edited works (Longley *et al*. 2001). Other good introductions include Chrisman (1997), Heywood *et al*. (1998) and Martin (1996a).

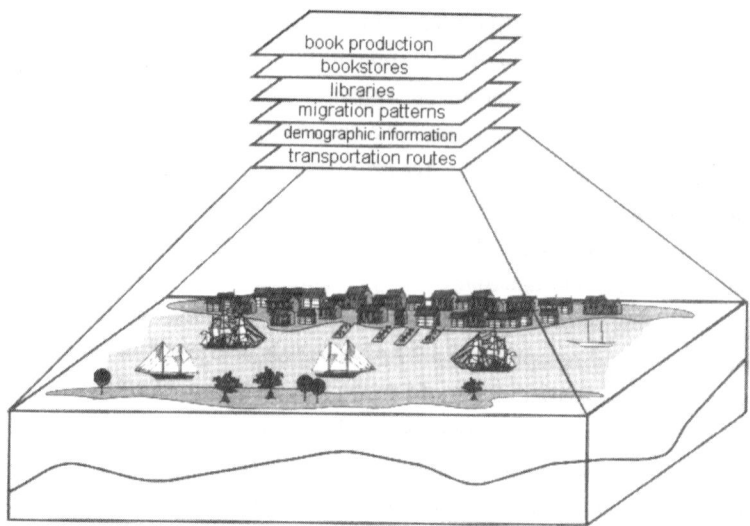

Figure 2: *Abstracting the real world into layers for the study of book history. Source: MacDonald and Black 2000, 510.*

As described in Section 1, GIS software uses a conventional database, termed *attribute data* (sometimes termed 'non-spatial data'). Each record in the attribute data is given a spatial reference using coordinates: these coordinates are termed the *spatial data*. There are two main types of GIS data model: in *vector data models* the spatial data consist of *points*, *lines* or *polygons*; while in *raster data models* the study area is usually sub-divided into square *pixels*, although other regular *tessellations* such as hexagons or triangles can also be used. In a vector model, therefore, space is sub-divided into discrete features, while the raster model attempts to represent space as a continuous *surface*. There are also other ways of sub-dividing the world into surfaces such as *triangular irregular networks* (TINs), often used to represent relief through *digital terrain models* (DTMs, also known as digital elevation models or DEMs).

Rather than store all their information about a study area in a single structure, the GIS separates the world into different *layers* (or coverages or themes) with each layer representing a different type of information. Most GIS texts give examples such as data on relief being one layer, data on the road network another, data on settlements another, data on rivers another, and so on. MacDonald and Black (2000) give an historical example. They are interested in exploring the development of print culture in the 19th century. They argue that the spread of books was partly due to factors directly related to books themselves, including sources of production such as printers and sources of supply such as libraries and bookshops. In addition the spread of books was also influenced by broader factors such as the transport network and demographic factors such as migration. Figure 2 shows how they can abstract the real world into layers in order to create a GIS that will allow them to explore book culture.

Understanding that GIS models the world through spatial and attribute data, that spatial data is made up of a small number of crude *graphic primitives*, and that data on different themes are brought together using layers is critical to an understanding of GIS. This basic model fundamentally defines what can and cannot be done within a GIS and the strengths and weaknesses of all GIS operations. There is an alternative approach to modelling the world in GIS using an *object-orientated* approach. Here features are not subdivided into layers but are instead grouped into classes and hierarchies. Although there are theoretical advantages to doing this, few commercially available GIS software packages have fully implemented this approach. For this reason object-orientated models will not be discussed in this guide.

2.2 ATTRIBUTE DATA

Attribute data are data in the form that most people understand by the term. Many GIS software packages include their own attribute databases but allow the user to link to external *database management systems (DBMS)* or spreadsheets. Examples of this include MapInfo, allowing the user to link to data in Microsoft Excel, ArcView to DBase, and ArcInfo to Oracle. Attribute data are frequently either statistical or textual. As the software improves and becomes more flexible, they can be in virtually any format that the DBMS used to store them can support. Increasingly this includes image formats, animations, hyperlinks, multimedia, and so on.

Most of the DBMSs used in GIS are *relational database management systems*. This means that two or more tables can be joined together based on a common field known as a *key*. With historical data this can often be either a place name or an ID number (note that place names are not considered to be spatial data: to be spatial the data must have a coordinate-based location), and it allows data from various sources to be integrated without requiring spatial data. For example, a user has a table of Poor Law data organised by Poor Law Unions (these were a type of administrative unit used in England and Wales in the 19th and early 20th centuries to administer relief of the poor), a table of voting statistics organised by parliamentary constituency, and some employment statistics based on towns. A *relation join* will join all three tables together and all of the data for 'Bristol', for example, will appear on a single row.

There are three main problems with doing this: firstly, the join has no knowledge that the entity referred to as 'Bristol' may be a different entity in each table. Secondly, problems will occur with names such as 'Whitchurch', which is not unique and appears in both Hampshire and Shropshire. Different software will handle this in different ways, the most common (and theoretically sound) being to duplicate rows of data. One way round this is to use more than one column as the key, for example, place name and county. The third problem with using place names is that their spellings must be identical to produce a match. Even minor differences in the use of hyphens or apostrophes will cause a non-match. This can be worked around using gazetteers that standardise all possible spellings and create a single spelling from an authority list, or through the use of ID numbers. Creating these can be time consuming.

Many attribute databases use *Structured Query Language* (SQL) to allow flexible querying and joining. This is often implemented though a Graphical User Interface but follows the basic structure:

    select <column names>
    from <table names>
    where <condition>

So, for example, we have two tables: 'unemp' that contains data on unemployment rates, and 'inf_mort' that contains data on infant mortality. These have the following fields:

Unemp	
Field Name	**Contents**
Name	The area's place name
U_rate	The area's unemployment rate
Tot_pop	The area's total population

Inf_mort	
Field Name	**Contents**
Name	The area's place name
Im_rate	The area's infant mortality rate
Births	The number of births in the area

Table 1: Sample tables of attribute data

The SQL query:

    select name, u_rate, tot_pop
    from unemp
    where u_rate>10.0

will select the names, unemployment rates, and total populations from the table unemp for places with an unemployment rate of over 10% as is shown in Table 2.

Name	U_rate	Tot_pop
Bolton	13.2	10,000
Oldham	12.1	7,500
Rochdale	10.9	8,000

Table 2: Sample data returned by the query above.

Relational joins are also implemented in this way. For example the query:

> select unemp.name, unemp.u_rate, inf_mort.im_rate
> from unemp, inf_mort
> where unemp.name=inf_mort.name

will select the unemployment rates from unemp and infant mortality rates from inf_mort where the values in the name fields in both tables are identical.

Unemp.name	Unemp. U_rate	Inf_mort. im_rate
Bolton	13.2	120
Oldham	12.1	115
Rochdale	10.9	106
Burnley	9.4	98
Colne	9.1	89

Table 3: Sample data returned by the query above

While relational databases and SQL are not fundamental to an understanding of GIS, knowledge of them can be useful in order to enhance an understanding of GIS data and GIS software. Many guides to the use of SQL and relational databases are available: see the bibliography for further information.

2.3 VECTOR SYSTEMS

In vector systems spatial data are represented either by points, lines, or polygons. A point is represented using a single coordinate pair. A line (or arc or segment) is represented by a string of coordinate pairs giving the start and end point of the line and the coordinates of all points where the line changes direction. A line's start and end points are often referred to as *nodes*. Polygons are created by completely enclosing an area by one or more lines. A basic polygon data model is shown in Figure 3. How lines are connected to create this model is known as *topology*. Both polygons and lines need topological information. A polygon needs to know the ID numbers of all the line segments that make up its boundaries, while a line segment usually knows which polygons are to its left and its right. For example in Figure 3, polygon p3 is bounded by line segments s4, s10, s11, s12, and s2. Line segment s10 has polygon p3 to its left and polygon p4 to its right.

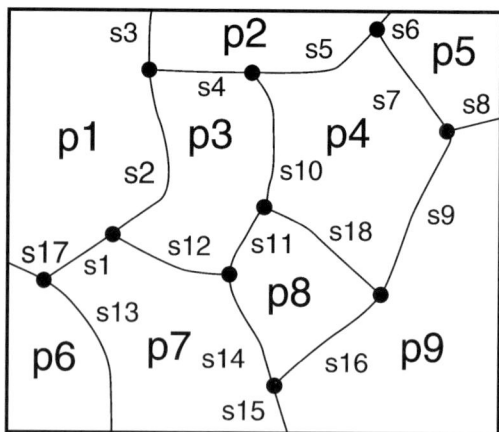

Polygon id	Line segment ids
p3	s2, s12, s11, s10, s4
p4	s7, s5, s10, s18, s9
p8	s11, s14, s16, s18
...	...

Segment id	Line segment coordinates
s1	(x1, x1) (x2, y2) (x3, y3)
s2	(x7, y7) (x8, y8)
s3	(x10, y10) (x11, y11)
...	...

Figure 3: Simplified structure of polygons in a vector GIS. The spatial data are created by grouping coordinate pairs into line segments. Polygons are then created by completely enclosing an area with one or more line segments. Modified from Jones 1997, 45.

Topology is essential if polygons are used, as without it boundaries would have to be stored twice: for example, line segment s10 would have to be stored as part of polygon p3's boundary and also as part of polygon p4's. It is important to note that the topological data model means that polygons cannot overlap and that every location in the study area can only belong to one polygon. Topology can also be used to turn a collection of lines into a *network*. This is usually based on each node knowing which line segments it is connected to, so in other words the nodes represent junctions. This can be extremely useful as it means that transport and other networks can be used, for example, to determine the shortest path between two points.

Attribute data is linked to points, lines, and polygons using a relational join on ID numbers as is shown in Figure 4. In modern software the creation and maintenance of topology, its associated ID numbers, and the links between the spatial and attribute data are kept hidden from the user. Nevertheless it is important to have some understanding of how the data are assembled.

Some features, such as towns or buildings, can be represented by either a polygon or a point. The choice of which representation to use is dictated by the purpose they are to be used for and, in particular, the scale of use. Polygons can be represented by points if required, usually by using the centre point of the polygon, termed the *centroid*.

Spatial data

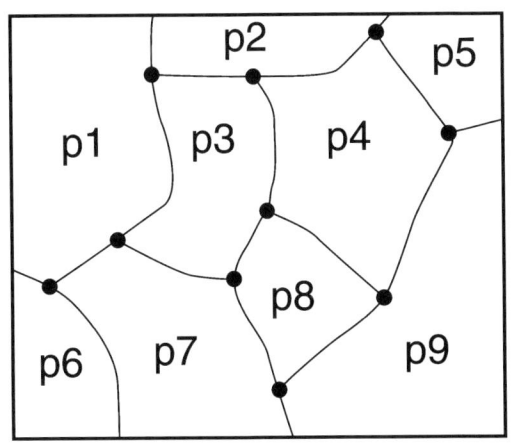

Attribute data

Polygon id	Parcel no.	Owner	Value (£)
p1	856	F. Hollins	1356
p2	455	D. Newson	2459
p3	152	L.U. Doyne	1754
p4	357	F. Futter	1269
p5	358	F. Futter	2098
p6	480	R.F. Mutter	3381
p7	390	O.R. Doyne	2862
p8	840	P. Gween	3750
p9	362	F. Futter	1089

Figure 4: Linking spatial and attribute data. The attribute data are stored in a conventional database table and linked to the spatial data using polygon ID numbers. Point and line data are linked to attributes in the same way. Modified from Jones 1997, 45.

The Great Britain Historical GIS (Gregory and Southall 1998; 2000), SECOS (Gatley and Ell 2000) and the Great American History Machine (Miller and Modell 1988) all have attribute databases containing a variety of census and other data. These are linked to polygon representations of the administrative units used to publish the data such as registration districts in England and Wales, and counties in the United States. In this way the data are put on the map and the map is populated with data.

2.4 RASTER SYSTEMS

The *raster data model* sub-divides *space* into square *pixels* or other regular *tessellations* to provide a continuous representation of the study area rather than subdividing it into discrete points, lines, or polygons. For example, to represent relief each pixel might have its height as an attribute; to represent land-use each pixel would have a land-use class attached to it, and so on. Although less likely to be of use to historians, *satellite images* are a more complex form of raster data. On the image the earth's *surface* is sub-divided into pixels with each pixel storing information about the amount and type of light being reflected by that part of the earth's surface, for example, how much green, how much blue, how much red, and how much infra-red.

The simplest raster file formats are two-dimensional arrays in which each value corresponds to a pixel on the grid that the array is modelling. The header information provides data on, for example, pixel size and the location of the bottom left-hand corner of the grid, while the remainder of the file simply consists of the values for each cell. Figure 5 shows an example of this. The original map shows three types of land-use: agriculture, urban and forest. The raster sub-divides the study area into pixels with each pixel being allocated a numeric value

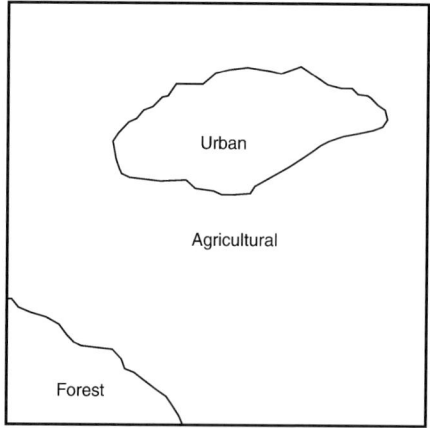 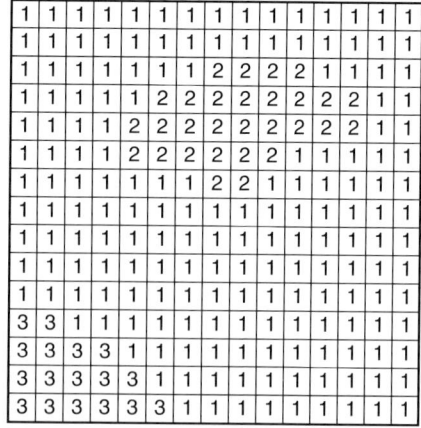

Figure 5: The raster data model. The source map of the study area shows three land-use classes: agricultural, urban and forested. The raster data model sub-divides the study area into pixels with each pixel being allocated to the land-use that covers the highest proportion of its area. Each pixel is given the number 1, 2 or 3 representing agriculture, urban or forest respectively.

representing land-use class. The pixel is allocated to the type of land-use that covers most of its area. Obviously the choice of pixel size is very important to the model. Too large a size will lead to a poor representation of the features, too small a size will lead to unwieldy file sizes. More sophisticated methods, such as *run length encoding* and *quadtrees* for example, can be used to compress the file sizes, but the basic model remains the same. For more information on the details of different raster structures see, for example, Section 3 of Heywood *et al.* (1998).

In general, raster data are more suited to environmental applications while vector data are more suited to human activity. Raster systems model complex spatial patterns with limited attributes, such as land-use patterns, very well, while the vector data model is better for more clearly defined space with complex attributes such as census data. There are exceptions to this: Martin (1996b) uses derived raster surfaces to model 1981 and 1991 census data and compare change between the two, claiming that the raster model provides a more realistic model of the underlying population distribution. A good example of a raster system in a historical context is provided by Bartley and Campbell (1997). They examined the Inquisitions Post-Mortem of the 14th century and used these to create a raster GIS of medieval land-use that, they claim, is potentially the most detailed survey possible until the 19th century tithe surveys. A raster system was used because it provides a complete coverage of the land area, it provides a more realistic representation of land-use than polygons, and because it handles the inaccuracies of the sources better than polygons.

2.5 OTHER SYSTEMS: TERRAIN MODELLING WITH TINS

A *triangular irregular network* (TIN) can be created from data that, in addition to x and y coordinate values, also have a z value that usually represents altitude. Spot-heights are the most

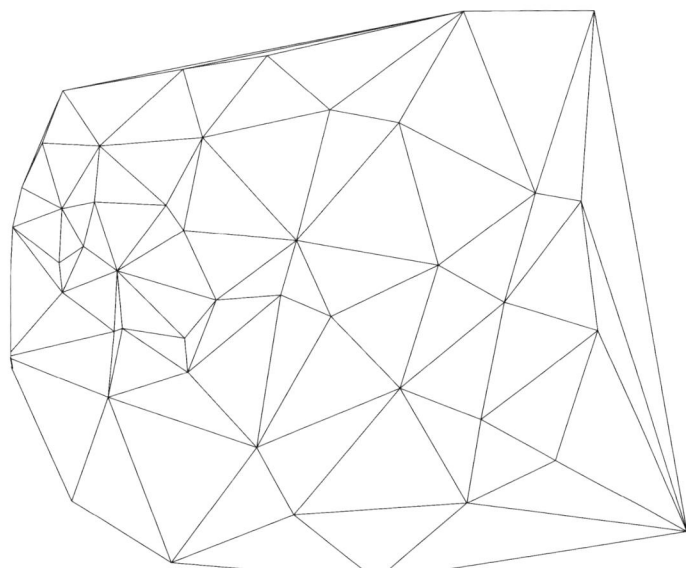

Figure 6: A Triangulated Irregular Network (TIN). From a series of points such as spot-heights, non-intersecting lines are drawn to the points' nearest neighbours. Using the 'heights' the gradients of the lines can be calculated and a virtual landscape (DTM) created.

common source of data in this form. The first stage in generating a TIN is to plot all the x,y coordinates as points. Next a straight line is drawn between each point and its nearest neighbours. This is shown in Figure 6. As the heights of all the points are known, the gradient of the line of these lines is also known. If data are available from enough points, these can be used to generate a surprisingly realistic, apparently three-dimensional, representation of the shape of the landscape, known as a *digital terrain model* (DTM). Often raster and vector data are *draped* over the DTM to provide features on the landscape. Models such as these are often shown on news programmes; for example, the siege of Sarajevo was regularly pictured this way.

Harris (2000) uses a terrain model to recreate the ancient landscape around an Adena Indian burial mound in Moundsville, West Virginia.

2.6 BRINGING IT ALL TOGETHER WITH LAYERS

Each separate theme of information is represented as a layer in the GIS. To build up a realistic representation of the study area a variety of layers are usually used. For example, a DTM may be used to model the relief of an area, a raster surface to model its land-use, a point layer used to represent buildings of interest, a line layer to represent rivers, a network layer to represent the transport system, and polygon layers to represent field patterns and administrative boundaries. All of these contain the relevant attributes. Exactly when to use a separate layer is often a matter of choice. For example, if we have three types of buildings – churches, hospitals, and museums – these could be stored as three separate layers, or they could be stored on a single layer with attributes to say what type of building each point represents.

Most common GIS software packages can handle most of these types of data although the majority will concentrate on one. *MapInfo* and *ArcView* are both primarily vector systems that have limited raster and terrain modelling functionality. *Spans* and *Idrisi* on the other hand are primarily raster based.

A good example of a study area modelled through a GIS is provided by Pearson and Collier's work on the parish of Newport in Pembrokeshire (Pearson and Collier 1998). They were interested in land ownership and agricultural productivity in the mid-19th century. To investigate this they needed to combine environmental data such as soils, slope, aspect and altitude, with historical statistics such as census data, and information from the tithe survey, a detailed and comprehensive survey of land ownership and land-use carried out in England and Wales during the early 19th century. Information from the tithe surveys was recorded as polygons. Each field was represented as a polygon with attribute information such as the owner, the occupant, the field name, the state of cultivation, the acreage and the tithe rent-charge. The environmental data started as altitudes for a regular grid of points 50m apart covering an area of 10km by 10km. This was used to create both a raster grid for which each cell had an altitude, and a terrain model that allowed slope and aspects to be calculated. Bringing the data together in this way allowed a detailed reconstruction of the various factors affecting agricultural productivity in the mid-19th century, and provided a starting point for more sophisticated analyses.

A second example is provided by Siebert (2000). He was interested in reconstructing the history of Tokyo's development with a particular emphasis on physical features such as the shoreline and rivers, the administrative areas, the population distribution and the rail network. To do this he created a variety of layers for each type of feature: the rail network was represented by line layers; the shoreline and the rivers were represented either using lines (for small rivers) or polygons (for larger ones); and the administrative areas were clearly represented as polygons, as were the census data. As with Pearson and Collier's work, bringing the data together in this way allows a comprehensive picture of the study area to be built up.

2.7 CONCLUSIONS

The core idea behind GIS is that it attempts to represent features that are, or were, on the earth's surface. The first stage in doing this is to take real-world themes that are relevant to the research and abstract them as layers, with each theme being represented by a separate layer. Features on each layer are represented using a combination of spatial and attribute data with each layer's spatial data consisting of one of four types of graphic primitives: points, lines, polygons or pixels. The resulting data are therefore highly abstracted from the real world. Where the data can be effectively represented using this model, GIS provides a powerful research tool because of its unique ability to handle location and attribute simultaneously. However, this structure fundamentally restricts the way that GIS can be used and where features cannot be effectively represented in this way the use of GIS may not be appropriate.

The implementation of most of what has been described in this Section is performed automatically by modern GIS software and is well hidden from the user, but it is important to understand the way in which GIS represents the world. Many datasets simply do not fit this model comfortably. If they don't, it is unlikely to be worth investing much effort in GIS. Where they do, this model represents a powerful way of integrating space into an analysis or exploration of a particular location, be it an area of only a few square metres, a whole country, or even the whole world. In addition to the primitive method of representing space, many other issues also need to be borne in mind. These include scale, accuracy and uncertainty. The GIS research community expends large amounts of energy worrying that GIS users will not understand these limitations. They should, however, be familiar to historians as they are the limitations of the original source plus the additional limitations created by converting the source into digital form. The problem is that for spatially referenced data these are more complicated than they may appear at first sight. The issues involved will be discussed in more detail in subsequent Sections.

Section 3: Acquiring Spatial Data

3.1 INTRODUCTION

There is no cheap and easy way of getting spatial data into a GIS. In many cases the *data capture* phase of a GIS-based project is the longest and most expensive of all phases. Bernhardsen (1999) suggests that the collection and maintenance of data accounts for 60 to 80% of the total cost, in terms of both money and time, of a fully operational GIS. Other authors give higher figures. Siebert (2000), for example, gives an honest account of the time taken and problems encountered building his historical GIS for Tokyo.

There are two basic sources of spatial data: *primary sources* where the data can be captured directly into the GIS, for example through the use of *Global Positioning Systems* (GPS); or *remote sensing* from satellites. More common for historians are secondary sources capture, where data from paper maps are converted into digital form. There are two ways of doing this: *scanning* the maps to produce *raster data*; or *digitising* the maps where points or line features are traced either directly from paper maps or from scanned images of the maps. This produces *vector data*. It is worth noting that the GIS definitions of primary and secondary data sources are different from the definitions used by historians. From a GIS perspective, primary data are data captured directly from the real world while secondary data are captured from abstracted sources such as maps. A map produced in the 1850s is a primary source from a historical point of view, but a secondary source from a GIS perspective.

An alternative to capturing data yourself is to acquire digital data from someone else. This is less time consuming and less risky but spatial data are often expensive, may have serious copyright restrictions placed on their use, and need to be fit for the purpose that you require. In the context of historical research, many datasets that may be required have not been digitised owing to a lack of demand for them, so there is frequently no alternative but to capture them yourself.

The first, and probably most important, decision to take when acquiring digital spatial data is what the source should be, and this applies whether the data are to be captured in-house or acquired from others. The limitations of the original source inherently limit any subsequent use of those data. The scale of the source is of particular importance here. In general, larger scale data are more flexible than smaller scale but will be more expensive, either in terms of purchase price or the time taken to capture them, and will have more redundant information which can lead to problems with file sizes. A second issue to consider is whether the reasons for the production of the original maps are compatible with the objectives of any digital representation.

3.2 SCANNING MAPS TO PRODUCE RASTER DATA

Scanning is a relatively straightforward process in which paper maps are placed on a scanner and a raster copy is produced. Smaller scanners are often relatively inexpensive; however, larger ones suitable for large map sheets are still expensive. The spatial resolution of the scanner, usually expressed in dots per square inch (dpi), and its spectral resolution (the number of colours it can distinguish), must be borne in mind, as this will affect the accuracy of the resulting data.

Converting from the resulting scan to the type of raster data described in Section 2 will require a certain amount of post-processing. If the source maps are relatively simple and focused on a single theme, such as land-use or soil type, this may be quite straightforward, whereas converting from more complex sources can be very time consuming.

3.3 DIGITISING MAPS TO PRODUCE VECTOR DATA

While scanning produces a copy of the source map, digitising extracts certain features from the source and creates point, line, and/or polygon layers from them. From an early series of Ordnance Survey (OS) inch-to-the-mile maps, a user might simply want to extract a layer of points representing the locations of churches. Other layers, such as the road and rail networks, administrative boundaries, and so on may also be extracted. However, it is not possible to create a direct copy of the source map in the way that scanning does.

Points are digitised by clicking on the features that are required. Lines are digitised by tracing along the lines and clicking at points where there is a significant change in the line's direction. Polygons are created by creating a topological structure on top of line data, as described in Section 2. Digitising can be done either directly from the paper map, using a *digitising tablet* or *table*, or by first scanning the map and digitising on screen, known as *head-up digitising*. Digitising tablets and tables (tables are usually larger) consist of a flat surface on which the map is firmly stuck down. The tablet or table's surface has a fine mesh of copper wires underneath it. There is also a *puck*, a hand-held device with a fine cross-hair and one or more buttons. To capture a point the cross-hairs are placed over the feature of interest and a button is pressed. The tablet or table is able to calculate the exact location of the cross-hairs from this and the coordinate is passed to the GIS software. With head-up digitising, a cursor is placed over the feature on the screen using the computer's mouse and a button is clicked to determine its exact location. Head-up digitising has the advantage that it creates a scanned copy of the source in addition to the vector data extracted from it. This both preserves a copy of the source, and can be used as a *backcloth* that provides context for the extracted vector features.

Digitising accuracy is extremely important especially where topology is to be created. A node will often have to be digitised two or more times to represent the end point of one line segment and the start point of another. Most GIS software provides tolerances that mean that if two nodes are within a set distance they will be 'snapped' together to form a single node. If the tolerance is set too high inappropriate features will be snapped. If it is too low then gaps will appear, known as *dangling nodes*, where polygons will not close properly. This will lead to corrupted topology.

3.4 GEO-REFERENCING

Whether data have been scanned or digitised, their underlying coordinate scheme at this stage of the data capture process will usually be in inches or centimetres measured from the bottom left-hand corner of the scanner or digitiser. *Geo-referencing* is the process by which these coordinates are converted into real-world coordinates on a *projection system*. This allows distances and areas to be calculated and data from different sources to be integrated.

Map projections are intricate and complicated. In a country such as Britain where the use of the National Grid is almost universal, a detailed understanding of projection systems is rarely necessary. In this guide only the briefest description will be given; further details can be found in works listed in the bibliography. The earth is a globe and locations on that globe are described using *latitude*, the number of degrees north or south of the equator, and *longitude*, the number of degrees east or west of the Greenwich meridian. Maps are flat sheets of paper. Projections are the translations used to convert from a curved earth to a flat map surface. Doing this involves distorting one or more of distances, angles, areas or shapes. A projection will also often convert from degrees of longitude and latitude to miles or kilometres from a particular location with longitude becoming the x-coordinate and latitude becoming the y-coordinate.

The British National Grid is a Transverse Mercator projection. The origin of its ellipsoid runs north-south at 2° west of the Greenwich meridian, approximating to the central spine of Britain. As one moves east or west from this line, distances in particular become increasingly distorted. Fortunately, as Britain is a long, thin country running approximately north-south these distortions are rarely significant at the kind of scales at which historians operate. Traditionally, Britain was subdivided into grid squares with sides of 100km. Each square was given a two-letter identifier and locations were expressed as distances in kilometres or metres from the south-west corner of the grid square. For example, NN is a grid square in the southern highlands of Scotland. Within a computer, this use of letter codes is clumsy. Instead a false origin is given for the whole country. This is a point south-west of the Scilly Isles that allows all locations on mainland Britain to be expressed to the nearest metre as non-negative integers of no more than six figures. Location 253000, 720000 is 253km east of the false origin and 720km north and is in the southern highlands. This structure allows easy calculations of the distances between any two points expressed using six-figure National Grid references.

Most GIS software packages make the process of geo-referencing appear quite straightforward. Using the source map the user finds the real world coordinates of a number of *reference points*, usually four. These are also called *tic points* and are often the corners of the map sheet. These points are then digitised. The user is then prompted to type in the real-world coordinates of the reference points and the software uses these four points to convert every coordinate in the *layer* to real-world coordinates. Frequently the software will also prompt for a projection system at this stage and convert the layer accordingly. This means that the locations of all points on the layer will be expressed in National Grid coordinates or whatever other coordinate system has been selected, and that all distances measured on the layer will be in metres, kilometres, or whatever referencing unit is used.

While this is relatively easy using modern maps where coordinate grids are shown and projection information is readily available, it can be difficult using older maps. Where no coordinate grid is provided, this can be worked around by finding reference points on the source map that are also mapped on modern maps (appropriate features may include churches, lighthouses, trig-points or railway stations). The coordinates of these points can then be found using modern maps that do include a coordinate grid. For the sake of accuracy, the modern maps used should preferably be larger scale than the source map. If no information is provided on map projections then books such as Delano-Smith and Kain (1999), Harley (1975), Oliver (1993), Owen and Pilbeam (1992) are good sources of information. For work on small study areas, the impact of projections will be so small that they may not be worth bothering with.

3.5 ERROR AND ACCURACY

As was previously stated, a digital representation of a paper map is at best of equal quality to the original map, but it will almost inevitably accrue some additional error or inaccuracy. It is important to distinguish between the different types of error and inaccuracy. Unwin uses a six-way classification as follows: *error* is the difference between reality and the digital representation of it; *blunders* are simply mistakes; *accuracy* is the closeness of results, computation or estimates to values accepted as true; *precision* is the number of decimal places given in a measurement which is usually far more than it can support; *quality* is the fitness for purpose of the data; and *uncertainty* measures the degree of doubt or distrust when using the data (Unwin 1995).

Scanning, digitising and geo-referencing are particular sources of locational error. To digitise a map it must first be placed completely flat on the digitiser or scanner. Even this is not always as easy as it sounds: maps may have been stored folded, paper warps over time, and so on. The accuracy of the scanning or digitising equipment itself is the next possible source of error, although if specialised (normally expensive) hardware is used this is usually only minimal. If head-up digitising is used there will be cumulative error as there is the error introduced by the scanning, and then error introduced by the digitising.

The next source of error comes from the user's involvement with the data capture process. With digitising, the person capturing the data has to place the puck or cursor over the point to be captured. Even a highly motivated and alert person will make minor positional errors. When digitising is done for many hours by low-paid staff, the potential for both inaccuracy and blunders is increased. Even with the best will in the world it is not always possible to capture the exact location accurately. Point symbols are not always a precisely defined point: for example railway stations on an OS 1:50,000 map are represented by a circle that is nearly 2mm in diameter. Line features are even more difficult. Digitising a line relies on the operator capturing each point at which the line changes direction. For gentle curves, such as those on roads, rivers or contour lines, this is inevitably a subjective choice and no two operators digitising a line of this type will ever digitise exactly the same points to describe it.

A final source of error is the geo-referencing process. Coordinates measured from a map will have a certain amount of error in them. The locations of the reference points will also have some error. This means that the placement of every location on the layer will be slightly distorted. Most software packages will provide a measure of the error expressed as *Root Mean Square* (RMS) error. This is often expressed in both digitiser units and also real-world units. It is recommended that the RMS error should not exceed 0.003 digitiser inches (ESRI 1994a). Even this standard is not always possible with historic maps, and it is important that minimum standards of accuracy are established as part of the data capture process and that these are documented (see Section 9). The key point to this section is that there are many issues associated with error and uncertainty in spatial data that may not be familiar to the historian. The problems these cause may seem daunting; however by facing up to the issues and being aware of their implications these should not cause serious difficulties to the research process.

3.6 DIGITISING ATTRIBUTE DATA

This is usually a more straightforward process than capturing spatial data. Attribute data are usually captured by scanning and Optical Character Recognition (OCR), or by typing. These can either be typed in directly against spatial features in the GIS software, or can be captured separately and joined to the spatial data using a relational join. For more information on good practice in capturing attribute data see Townsend *et al.* (1999).

3.7 RASTER-TO-VECTOR AND VECTOR-TO-RASTER DATA CONVERSION

The holy grail of spatial data capture is simply to be able to scan a map and then automatically extract point, line and polygon features. This is called *raster-to-vector conversion*, but it has not proved easy to implement. Some systems claim to be able to do this but the process usually requires a large amount of user interaction. Even then it can be inaccurate: for example, lines can pick up a stepped effect as a result of following the underlying raster grid pattern. *Vector-to-raster conversion* can be more successful, but care still needs to be taken. Many software packages include routines that allow both of these conversions to take place, but many also make exaggerated claims about the potential accuracy of the routines. Before committing to a major purchase, or major investment in time, it is well worth checking the results first rather than believing the sales pitch provided by the software company.

3.8 PRIMARY DATA SOURCES

Two main primary data sources are data from satellite imagery, and Global Positioning Systems (GPS). Satellite imagery is a form of raster data in which each pixel represents a part of the earth's surface. The exact dimensions of the pixel depends on both the type of scanner used and the post-processing applied, but is usually a square with sides of from 1 metre to around 100 metres. For each pixel some information is provided about the light that was reflected back from that part of the earth's surface when the image was captured. Again, the exact details depend on the type of scanner and the post-processing applied. From this basic information, more sophisticated knowledge can be developed on, for example, types of land-use, health of vegetation, and so on. It is not the intention to describe remotely sensed satellite imagery here, as many good guides are available. Obviously, satellite imagery has only become available in relatively recent times but it may provide useful information for historians; for example, in determining the location of certain features in remote areas, or to attempt to determine past land-uses. Aerial photographs may be an alternative (secondary) source of data on land-use in the past.

GPS receivers are an easy way of primary spatial data capture. The simplest form of GPS receivers give the coordinate for the current location of the user. This is calculated from a network of satellites launched by the United States military. Initially the accuracy of these locations was deliberately degraded for non-military users to around 100m. This has now stopped and accuracies of only a few metres are available to all users. More sophisticated systems allow the user to capture multiple points as the receiver is moved and download them directly into a computer. If more accuracy is required *differential GPS* (DGPS) is used. This requires the use of two receivers, one of which is kept stationary at a known location to assist in measuring the location of the roving receiver. This can potentially produce sub-metre accuracy.

3.9 BUYING DATA OR ACQUIRING IT FREE

For academics in the United Kingdom there are a variety of organisations that provide GIS data either free or at low cost. For historians the most useful of these is likely to be the History Data Service (HDS) <http://hds.essex.ac.uk/>, part of the Arts and Humanities Data Service (AHDS) <http://www.ahds.ac.uk/>. More detail on the role of the HDS is given in Section 9. EDINA <http://edina.ac.uk/> at the University of Edinburgh provides a suite of services including: UKBORDERS, which disseminates contemporary and historical boundary data for the UK, and Digimap, which provides a certain amount of Ordnance Survey (OS) digital data. The MIMAS <http://www.mimas.ac.uk/> service at the University of Manchester provides access to a variety of datasets, including Bartholomew's map data, while the KINDS project provides a variety of visualisation tools that enhance access to datasets. The Public Record Office (PRO) <http://www.pro.gov.uk/> is increasingly becoming involved in electronic publishing, although only a limited amount of data are currently available online.

Other sources of free or low-cost data include the Digital Chart of the World <http://www.maproom.psu.edu/dcw/>, which provides data on a variety of different countries; and the United States Geological Survey (USGS) <http://www.usgs.gov/>, which provides a variety of products often at relatively cheap rates. The Electronic Cultural Atlas Initiative (ECAI) <http://www.ecai.org/> has started to produce online e-publications of historical datasets, and also incorporates a metadata clearing-house that allows users to search for historical GIS datasets.

A variety of commercial companies sell spatially referenced data. The most obvious source of this in Britain is the Ordnance Survey <http://www.ordsvy.gov.uk/> which sells digital versions of many of their map products. Other commercial sources include the AA <http://www.theaa.com/aboutaa/data_sales.html> and Bartholomew's <http://www.bartholomewmaps.com/>. Purchasing these data can be expensive and copyright limitations may be placed on their use, but buying the data does provide high-quality products quickly and without the risks involved in capturing it yourself.

Whatever the source of data it is important to bear two things in mind: the first is the limitations of the original source material and its fitness for the purpose that you want to use it for. The second is any limitations that are placed on the use of the data through copyright or other restrictions. Misuse of this may result in civil or criminal action and is also likely to lead to people and organisations being less willing to disseminate data.

3.10 CONCLUSIONS

Acquiring spatial data is usually either time consuming or expensive or both. It is not to be entered into lightly. Once spatial data are available in digital form, however, they become a powerful resource with uses that may go far beyond the original reason why they were captured. It is important to remember that digital spatial data have all the limitations of the original source data, be they maps or other sources, plus some additional limitations and errors introduced by the data capture process. As will be described in Section 9, good documentation and *metadata* are essential elements of the data capture process, as this allows other users to evaluate the fitness of the data for their purpose. This ensures that data are widely used but only for suitable purposes.

Section 4: Basic GIS Functionality: querying, integrating and manipulating spatial data

4.1 INTRODUCTION

This Section introduces the core functionality of GIS. It describes the first steps that a user can undertake once they have their data in a GIS software package. This is the last Section that approaches GIS in this way; the remainder of the book focuses on issues rather than functionality. As described in Section 1, GIS software combines computer-mapping functionality that handles and displays spatial data, with database management system functionality to handle attribute data. This Section describes the basic tools that this provides to the user that are not available in other types of software packages. The basic functionality is as follows:

- *Querying* both spatially and through attribute
- Manipulating the spatial component of the data: for example, through changing projections, *rubber sheeting* to join adjacent layers of data together, and calculating basic statistics such as areas and perimeters of polygons
- *Buffering* where all locations lying within a set distance of a feature or set of features are identified
- Data integration, either informally by simply laying one layer over another, or formally through a mathematical *overlay* operation
- Areal interpolation.

4.2 QUERYING DATA

As with all database systems, one of the core parts of GIS functionality is the ability to query the data. With GIS software there are two basic forms of querying: spatial and attribute. *Spatial querying* asks the question 'what is at this location'? This is often done by simply clicking on a feature and then listing its attributes. More complex spatial queries could select all the features within a box or a polygon, or ask 'what is near to this feature'? These more complex queries often require the use of buffering or overlay techniques as described later in the Section. *Attribute querying* asks the question 'where does this occur'? If a user has a layer consisting of the locations of churches with some information about each church, an attribute query could select all the churches whose denomination is Catholic and then draw them with a certain symbol. The user could then query the database to select all Protestant churches and draw these with a different symbol to compare the patterns. There is much more on visualisation in Section 6; the purpose of introducing it here is simply to show how querying and mapping are inextricably linked.

4.3 MANIPULATING AND MEASURING SPATIAL DATA

Most GIS software packages come with a suite of options that allow the user to manipulate spatial data. One of the most basic of these is simply to change the projection system used. This can significantly alter the appearance of maps, particularly maps of the world, and can also make it possible to integrate data from layers that use different projection systems. In Britain, this could be used to take a variety of early maps on different projections and re-project them onto the National Grid. This would allow comparisons between different early maps, as well as with modern ones. Putting adjacent map sheets onto the same projection allows their digital representations to be joined to form a single layer. Where there are distortions to the sheets the maps will have to be *rubber-sheeted* (or *edge-matched*) to ensure that the edges of the two sheets make a perfect join. This involves telling the software where certain key points are on the layer and where they actually should be. The entire layer will then be distorted using these references. Figure 7 shows an example of this.

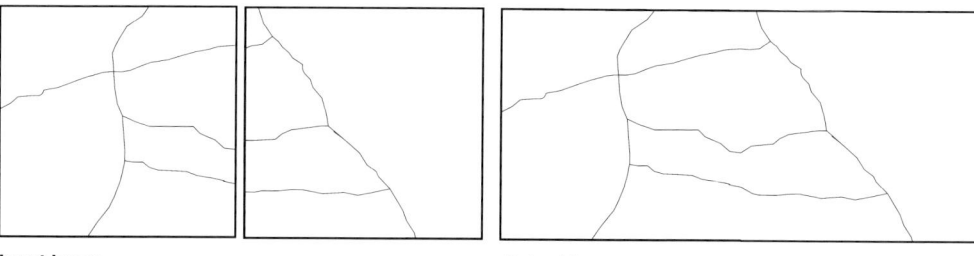

Input layers   **Output layer**

Figure 7: Rubber sheeting to join two layers together. A user wants to join the two input line layers together, but due to inaccuracy in one or both of the layers the lines do not match exactly. As part of the joining process the lines on the right-hand layer are systematically distorted to allow them to join to the corresponding line on the left. This distortion is at its maximum at the end points of the line and reduces as we move away from the join. The layers can then be seamlessly joined.

Most GIS software packages will also calculate basic statistics about their spatial features. Typical examples include calculating the length of lines, the area and perimeter of polygons, and the distances between points. There are many examples of why these basic measures can be useful. These include measuring distances along a transport network, the use of areas to calculate population densities, and calculating the distances between settlements.

4.4 BUFFERING, THIESSEN POLYGONS AND DISSOLVING

There are times when rather than simply being interested in the locations of a type of feature,

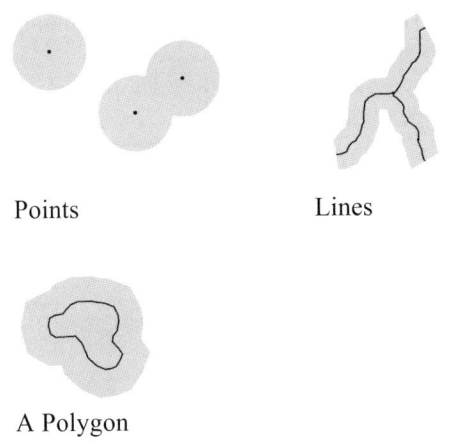

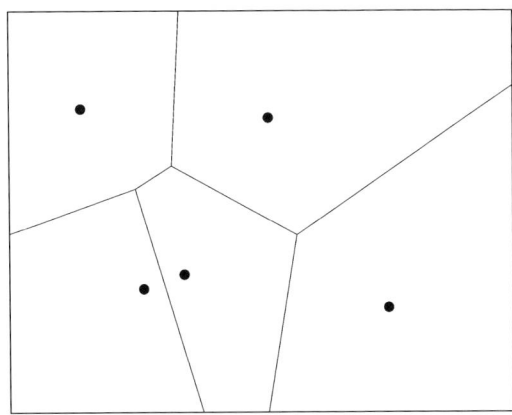

Figure 9: Thiessen polygons. The input layer is the set of points. From these, Thiessen polygons are defined in which each polygon defines the area closest to the point that lies within it. The boundary lines are thus lines that are equidistant from two points.

Figure 8: Buffers around points, lines and a polygon. In all cases the buffer polygons are shaded light grey while the input features are in black.

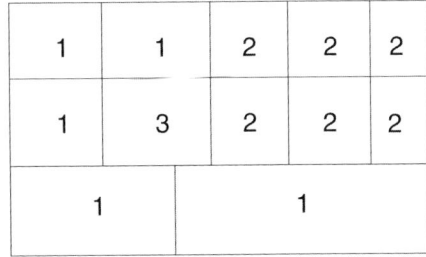

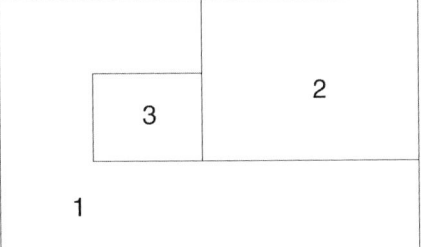

Input layer **Output layer**

Figure 10: Dissolving to aggregate polygons. The input layer is a polygon layer that could, for example, represent fields growing different types of crops. The user wants to merge these polygons to create a new layer of crop types. This is done with a dissolve operation that removes the unwanted internal boundaries creating the new output layer of aggregate polygons.

a user is interested in the locations within a set distance of a feature. Examples of this might include wanting to know all areas within (or outside of) 1km of a hospital, or areas within 10km of a railway line, or within 5km of an urban area. Where information of this type is required a *buffering* operation is used. Buffering takes a point, line, or polygon layer as input and produces a polygon layer as its output, as shown in Figure 8.

A user may also want to allocate catchment areas to a point dataset, and this can be done by generating *Thiessen polygons* (also known as Voronoi polygons). This creates a polygon layer in which the polygon boundaries are lines of equal distance between two points. This means that a polygon is the area that is nearest to the point that generated it, as is shown in Figure 9.

This is a simple form of *interpolation*, whereby data are allocated from one set of spatial units to another.

Another option occurs where a user wants to create aggregate polygons from a more detailed layer. For example, a user might have a polygon layer where each polygon represents a farmer's field with attribute data that includes crop type. If the user is only interested in where particular crops are grown, then many field boundaries represent redundant information that can be removed. This is done by what is called a *dissolve* operation whereby the boundaries of adjacent polygons with the same crop type are removed to form aggregate polygons. This is shown in Figure 10.

4.5 BRINGING DATA TOGETHER TO ACQUIRE KNOWLEDGE

Manipulating the spatial component of a single layer of data is useful, but the full potential of GIS lies in its ability to integrate data from a variety of layers. At a basic level this merely involves combining layers on-screen to compare patterns. This might be as simple as taking a raster scan of a map and placing a vector layer over the top. The raster layer provides a spatial context for the features in the vector layer. Another option is to lay one vector layer on top of another; for example to compare the pattern of roads with the location of farms to see which farms lie near the major roads. Field boundaries might be a third layer added to this. This approach goes beyond basic mapping, as querying the underlying attribute database allows a detailed understanding of a multi-faceted study area to be developed. In this way an integrated understanding of the problem can be derived from many (possibly highly disparate) sources.

Healey and Stamp provide an example of this in their study of regional economic growth in the north-eastern United States (Healey and Stamp 2000). They have created a large and comprehensive database that contains industrial plants, such as blast furnaces, foundries and coalmines, represented by points. This is combined with polygon data showing the boundaries of natural resource deposits and, in areas of more detailed study, land parcels. To study the economic development of these they also needed a database of the transport system, and this is reproduced by layers of lines that include the railroad and canal networks, the turnpike roads and the rivers. This can be combined with aggregate, background information, such as that extracted from both the population and industrial censuses. Pearson and Collier use a similar approach in their study of agricultural productivity (Pearson and Collier 1998). As was described in Section 2, they combine environmental information in the form of raster grids and terrain models with information from the tithe surveys and census data represented by polygon layers. Siebert brings together information on changing land-use, changing transport systems, administrative units, and population distribution to explore the urban development of Tokyo (Siebert 2000).

4.6 FORMALLY INTEGRATING DATA THROUGH OVERLAY

In addition simply to combining layers, querying them and comparing them, layers can be combined to produce new layers through geometric intersections. This is called *overlay*. Any of the three types of vector data can be overlaid with any of the others, as is shown in Figure

Section 4: Basic GIS Functionality

Input layer 1   **Input layer 2**   **Output layer**

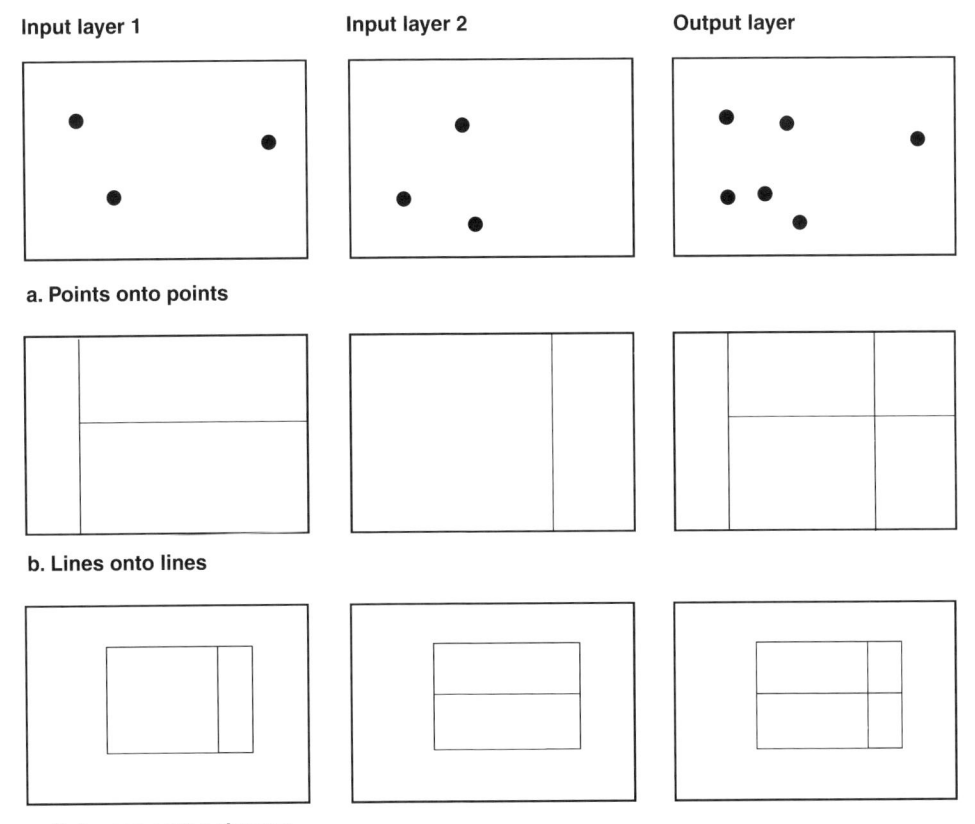

a. Points onto points

b. Lines onto lines

c. Polygons onto polygons

Figure 11: Different types of overlay operations. Examples of overlay operations combining point, line and polygon data. It is also possible to overlay, for example, point with polygon, line with polygon, etc.

11. An overlay operation combines not only the spatial data but also the attribute data. This has many potential uses. For example, a user has a polygon layer containing data about administrative units, such as Irish baronies, and wants to find out what proportion of each barony was covered by water using a polygon layer showing lakes. An overlay operation would produce a new polygon layer that combined the attributes of both polygon layers, as shown in Figure 12, and thus each new polygon would have both the barony attributes and the attributes from the water layer. It is also possible to combine point or line layers with polygon layers using overlay. For example, as inputs we might have a point layer representing towns and a polygon layer representing baronies and we want to determine which towns lay in which barony. Overlaying the two layers would produce a point layer with the barony polygon attributes added to each point, thus giving us the required information.

Combining buffering and overlay allows complex spatial queries and operations to be performed. For example, with a line layer showing the road network and a point layer containing farm locations, a user may want to calculate which farms lie within 1km of a major road. This

Spatial Data

Barony

```
        15
   17
16
      18
```

         88

    90

Output

```
     2    1
       3
    4
 5
    6   7
```

Attribute Data

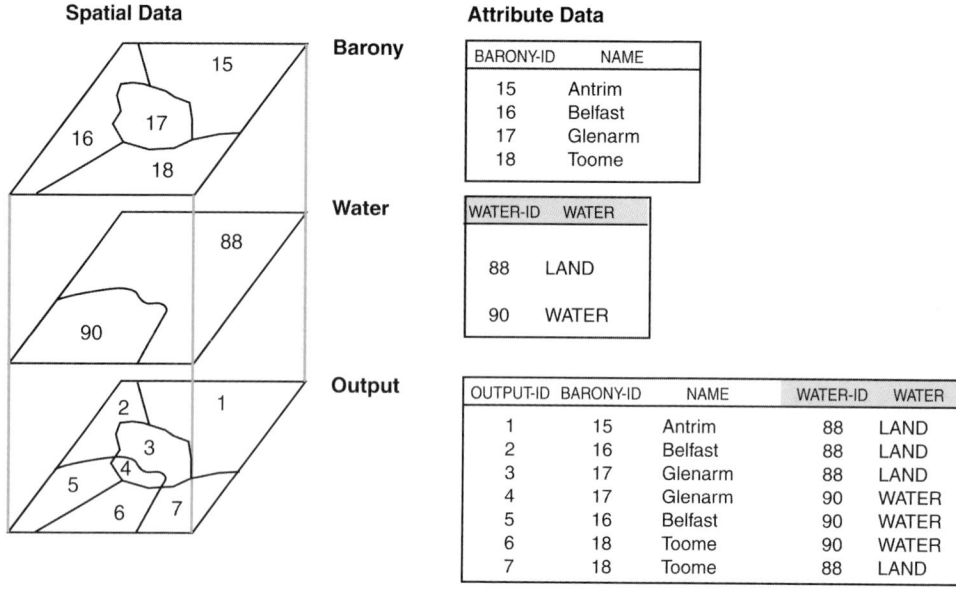

Barony

BARONY-ID	NAME
15	Antrim
16	Belfast
17	Glenarm
18	Toome

Water

WATER-ID	WATER
88	LAND
90	WATER

Output

OUTPUT-ID	BARONY-ID	NAME	WATER-ID	WATER
1	15	Antrim	88	LAND
2	16	Belfast	88	LAND
3	17	Glenarm	88	LAND
4	17	Glenarm	90	WATER
5	16	Belfast	90	WATER
6	18	Toome	90	WATER
7	18	Toome	88	LAND

Figure 12: Spatial and attribute data being combined using an overlay operation. The hypothetical barony and water layers are overlaid to produce the output layer. This new layer's attributes are the combined attributes of the two input layers. Modified from ESRI 1997, 8–18.

can be done as shown in Figure 13. First, the user selects only the major roads from the road layer and copies these to a new layer. A buffer is then placed around the new layer so that a polygon layer is created in which the polygons represent areas within 1km of a major road. If only farms within 1km of a major road are required, then a 'cookie cutter' overlay can be used. In this only farms on the input layer lying within a polygon on the buffer layer will be copied to the final layer. The final layer only contains five of the original farms and will contain all the attributes of both the farms and roads source layers.

Perhaps more than any other GIS operation, overlay tests the accuracy of the input layers. If a point layer is overlaid with a polygon layer then inaccurate polygon boundaries can easily lead to a point lying near a boundary being allocated to the wrong polygon. Accuracy is tested further where two polygon layers intersect. As was discussed in Section 3, it is unlikely that any two operators will digitise a curve in exactly the same way even if the same source map is used. If this happens then the overlay operation will lead to *sliver polygons* being formed. These are very small polygons formed in the manner shown in Figure 14. This may seem like a trivial problem but is in fact the bane of vector overlay operations. It is possible to attempt to remove sliver polygons automatically. They tend to be long and thin and thus have a small area compared to the length of their perimeter. While this can be used to identify slivers, deleting them can still be problematic as it requires a decision about which boundary should be deleted. The problems caused by sliver polygons will depend on the scale and accuracy of the two sources, and the accuracy of the digitising. If the two layers have both been digitised to a high standard of accuracy from high quality source maps of similar scales, then the problems are

Section 4: Basic GIS Functionality

Roads layer
Attributes: Road type

→ Select main roads →

Main roads layer
Attributes: Road type

↓ Buffer main roads

Buffer layer
Attributes: Road type

↓ Overlay farms and buffer

Farms layer
Attributes: Farmer's name

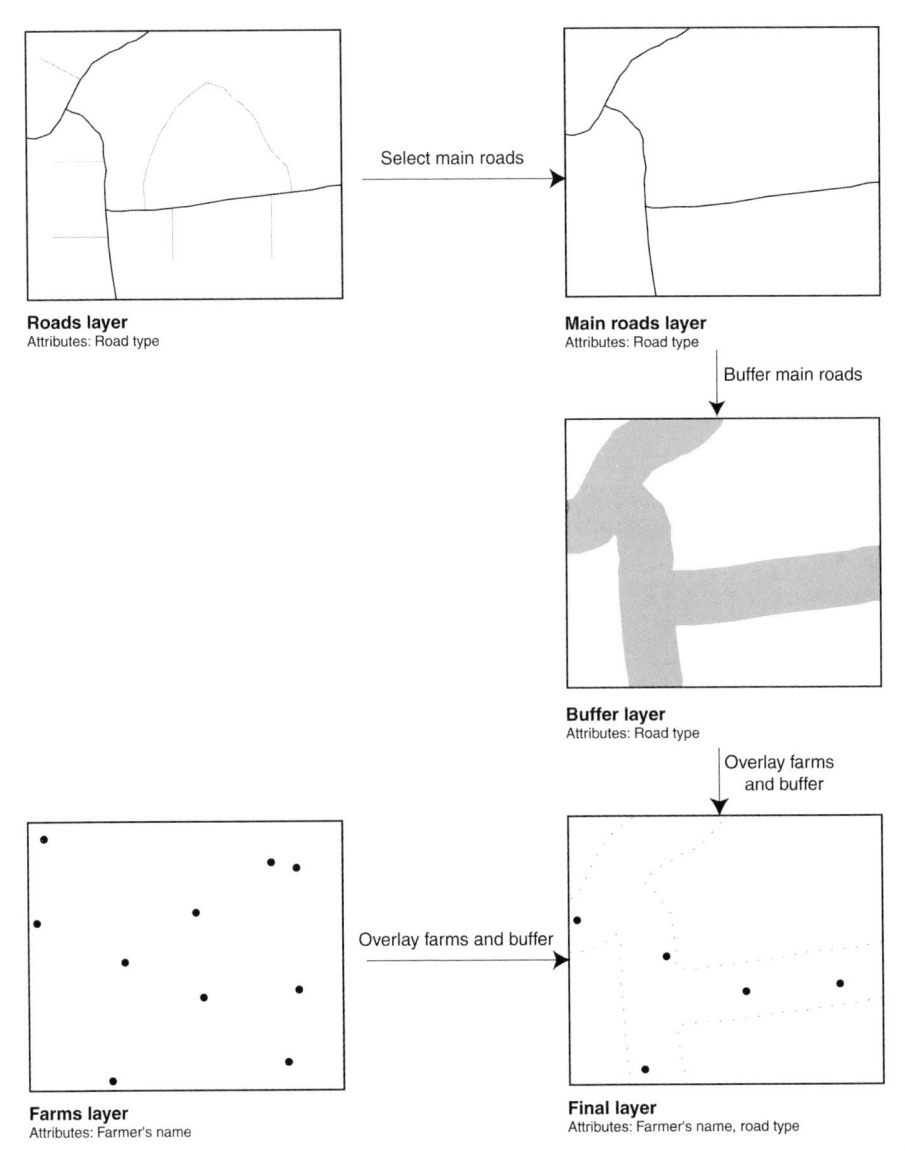

→ Overlay farms and buffer →

Final layer
Attributes: Farmer's name, road type

Figure 13: Spatial manipulation to solve problems. A user has two layers: a line layer showing roads in an area, and a point layer showing the location of farms. They want to select only those farms lying within a set distance of a main road. To do this they first select only the main roads from the roads layer using an attribute query. These are saved to a new main roads layer. A buffer is then placed round the main roads to create a buffer layer. In the final stage this buffer is overlaid with the farms layer using a 'cookie-cutter' overlay so that only points within the buffer are saved to the final layer. The final layer contains all the attributes from the source layers. Note that the outline of the buffer on this layer is shown for diagrammatic purposes only. The actual layer would consist only of points.

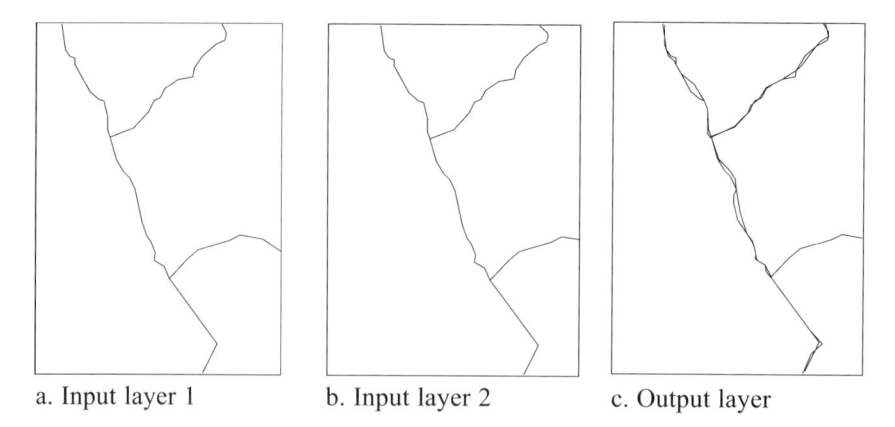

Figure 14: The creation of sliver polygons. a and b show a fragment of four polygons whose boundaries are supposed to be identical. When they are overlaid the minor differences between them result in the formation of small sliver polygons as shown in c.

Input 1

1	1	1	1	1
1	1	1	1	1
2	3	3	1	1
2	2	2	1	1
2	2	2	1	1

Input 2

0	0	0	2	2
0	1	3	2	2
0	1	3	2	2
1	1	2	2	2
1	1	1	2	2

Output

1	1	1	3	3
1	2	4	3	3
2	4	6	3	3
3	3	4	3	3
3	3	3	3	3

Figure 15: Map algebra with raster data. In raster overlay the values of cells in the output layer is calculated from the results of a mathematical operation on the input layers. In this example the two input layers have been added. Other operations such as multiplication and subtraction can also be used.

likely to be minimal and can usually be solved by automated procedures within the software. If any of these three criteria are not met there is likely to be a significant job tidying the resulting output layer.

Overlay can also be performed on raster datasets providing they use the same pixel sizes. This is sometimes referred to as *map algebra* as two or more input layers are used to create an output layer whose cell values are calculated based on a mathematical operation between the input layers. An example of this is shown in Figure 15 where cell values on the two input layers are added to calculate values on the output layer. Other mathematical operations such as subtraction and multiplication can also be used.

When two layers are combined using an overlay operation, the resulting layer will be at best as accurate as the less accurate layer. Unfortunately, the result is likely to be more inaccurate than this as error will be cumulatively added from both layers. This is termed *error propagation* and means that, as layers are combined, errors and uncertainty can multiply surprisingly quickly. This means that when multiple overlays are performed this must be done with the limitations of all the source layers being borne in mind.

An example of the use of overlay in historical research is provided by Lee (1996). The 19th century censuses of Ireland published data using baronies. These were relatively large spatial units and Lee wanted to estimate the internal population distribution of baronies in County Antrim to provide a more realistic representation of the population distribution. She believed that the distribution was likely to be affected by the presence or absence of large water features, altitude, and the proximity and function of nearby settlements. Her GIS consisted of:

- A polygon layer of the 16 baronies in County Antrim
- A polygon layer of large water features
- A polygon layer showing altitude (for example, land less than 200 feet, 200 to 400 feet, etc.)
- A point layer showing the location of settlements with attribute data concerning their size and function.

The barony, water feature, and altitude layers were overlaid to produce an output layer with 263 polygons. She then overlaid the settlement layer onto the centroids of these polygons so that distances from each settlement to each of the 263 centroids could be calculated. Finally, she used a rather arbitrary model to allocate the barony populations to each of the derived polygons based on the barony population, whether the polygon was covered in water, the polygon's altitude, and the distance from the polygon's centroid to nearby settlements. This shows how integrating a variety of disparate datasets can be used to generate new datasets.

4.7 INTEGRATING INCOMPATIBLE POLYGON DATA THROUGH AREAL INTERPOLATION

A final issue to be discussed in this Section is *areal interpolation*. This commonly occurs where there are two or more polygon layers of socio-economic data and the polygons represent administrative units, which a user wants to integrate. Where the two sets of polygons nest perfectly because the administrative units used were identical, this is a simple operation. Where they do not, for example if we are comparing census data published using registration districts with election data published for constituencies, then overlay does not provide the complete answer as it is uncertain how to allocate data to the resulting polygons. The traditional response to this is aggregation, which results in a loss of spatial detail, something that a GIS approach should attempt to avoid (see also Section 7).

Areal interpolation can be used instead. First we overlay the layer containing the input data onto the layer we want to estimate the populations for. These are termed the source and target layers respectively. The overlay generates the 'zones of intersection' between the two layers. The problem is then to estimate what proportion of the data from a source polygon to allocate to each zone of intersection. The simplest method of estimation is termed 'areal weighting'. This is shown in Figure 16. The source and target units are overlaid to form the zones of intersection M, N, O and P. These polygons have all the attributes of the source and target polygons plus the areas of the new polygons as calculated by the overlay. The final column of attribute data, 'Est Pop', is added by the user. Its values are estimated based on the area of the zone of intersection compared to the source polygon. Polygon N has an area of 30 while its source polygon, 1, had an area of 100. As a result 30% of polygon 1's population is allocated

Source

1
2

SID	Area	Pop
1	100	50
2	100	100

Target

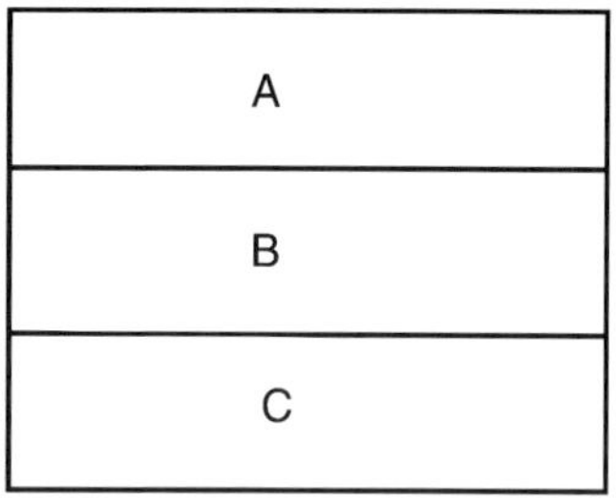

A
B
C

TID	Area
A	70
B	60
C	70

Intersection

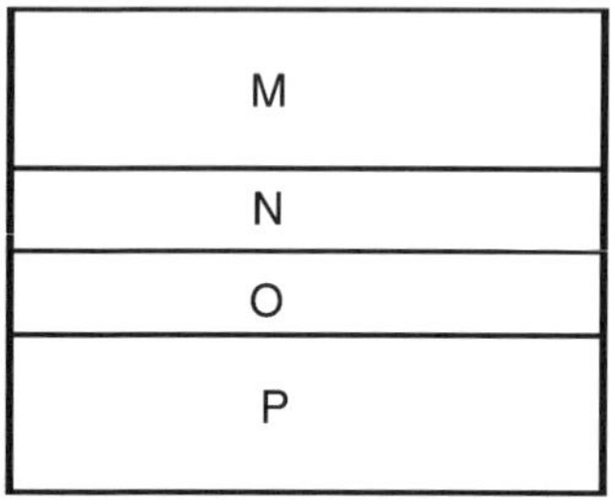

M
N
O
P

ID	SID	S Area	S Pop	TID	T Area	Area	Est Pop
M	1	100	50	A	70	70	35
N	1	100	50	B	60	30	15
O	2	100	100	B	60	30	33
P	2	100	100	C	70	70	67

Figure 16: Areal interpolation. We wish to allocate total population from a set of source units (1 and 2) onto a set of incompatible target units (A, B and C). The first stage is to perform an overlay. The new attribute data gives us the area and population of the source polygons, the area of the target polygons, and the areas of the new polygons calculated as part of the overlay. The population of the polygons on the intersection coverage can then be estimated by multiplying the population of the source polygon by the new polygons' areas and dividing by the area of the source polygon. For polygon N this is 50*30/100=15. The populations for the target polygons are then calculated by adding the values for the appropriate zones of intersection, thus A=35, B=15+33=48 and c=67.

to polygon N giving 15 people. The final stage is to aggregate the newly estimated data to target zone level so we estimate that target polygon 2 has a population of 15+33=48.

The assumption of even population distribution is obviously extremely unrealistic. For example, registration districts in England and Wales usually consisted of a market town and its hinterland, hardly a likely candidate for an even population density. Various techniques have been devised to work away from this assumption, mainly by using further knowledge that allows us to estimate where within the source zones the population is likely to be concentrated. This type of functionality is rarely properly incorporated into GIS software, and if it is then the technique and its limitations are likely to be hidden from the user. This means that users are likely to have to implement the appropriate procedure for themselves.

An example of the use of areal interpolation is provided by Gregory *et al*. (2000). They wanted to compare three quantitative indicators of poverty: infant mortality, overcrowded housing, and unemployment, as they changed in England and Wales from the late-19th century to the late 20th. To do this they compared data from four time periods: the late 19th century, the 1930s, the 1950s and the 1990s. All the data were available as polygon data from the census or *Registrar General's Decennial Supplements*, but used significantly different reporting geographies. The late 19th century data were published using approximately 630 registration districts, while the two dates from the mid-20th century used approximately 1,500 local government districts (however, even these were difficult to compare as the system was extensively reformed between the two dates). Modern data were available at much more spatially detailed levels, with as many as 100,000 units. To allow direct comparison, all the data were interpolated onto the least spatially detailed units, the 630 registration districts. This resulted in the loss of significant amounts of spatial detail from the later data and also introduced some error to the results, but it did enable a consistent time-series to be generated that allowed them to compare the changing patterns of inequality over time at a geographically consistent scale.

4.8 CONCLUSIONS: INFORMATION FROM SPATIALLY DETAILED, INTEGRATED DATABASES

GIS software provides extensive functionality that allows a user to approach his or her dataset in a way that combines the spatial and attribute components of their data. This functionality leads to added value being extracted from an existing dataset. All datasets have limitations, and the extra functionality provided by GIS software allows us to use the data in ways that their creator would never have envisaged. As a result it is important to consider the limitations of all layers when manipulating them with the GIS. It is also important to consider the limitations of the techniques used on the data, particularly those that integrate data. As long as the results of spatial operations are understood within these limitations, GIS software provides new functionality that should allow new understanding to be derived from spatially referenced data.

In this Section we have started to see the usefulness of the combined spatial and attribute data model used by GIS. This allows data to be queried and integrated in ways that no other approach can manage. The key advantage of this is that it allows the complexity of the data to be handled without undesirable simplification of the data.

Section 5: Time in Historical GIS

5.1 INTRODUCTION

For many years researchers in both geography and history have been arguing that to truly understand a phenomenon there must be a proper handling of both location and time (see for example Haggett 1979; Langton 1972; Massey 1999; Parkes and Thrift 1980; Thrift 1977). This effectively means that data should be handled using all three of their components: attribute, space, and time. As we have seen, GIS is good at handling attribute and space. Unfortunately most commercial GIS software packages do not include temporal functionality. This is because there are some serious conceptual issues that present barriers to handling time fully within the GIS data model. Users are thus left largely on their own in how they approach handling spatio-temporal data. This makes implementation difficult but provides the researcher with the opportunity to develop solutions that are sympathetic to both data and research, rather than being saddled with off-the-shelf solutions provided by vendors. This Section reviews the issues and problems associated with using time in GIS, and looks at some examples of how researchers have approached handling all three dimensions of data simultaneously.

5.2 THE NEED FOR UNDERSTANDING THROUGH SPACE AND TIME

Researchers have long argued that to truly understand the world, one must understand change through both time and space. This was the argument underlying Langton's article where he claims that rather than simply comparing isolated snapshots that are assumed to be in equilibrium (termed *synchronic analysis*), researchers should be able to study how processes operate through time "cutting across a successive series of synchronic pictures of the system" (Langton 1972, 137). He terms this *diachronic analysis*. More recently, Massey presents a strong argument for the need among geographers for a full understanding of space-time (Massey 1999). She argues that we need to be able to understand time to tell the story of how an individual place developed, and to understand space to understand the complexity of the way different places develop. Only by having multiple routes through space and time can the full complexity of the world be understood. Unfortunately, the complexity of handling data's three components simultaneously has usually led to researchers either simplifying space to preserve temporal detail, or simplifying time to preserve spatial detail. Cliff and Haggett (1996) summarise this by saying "If we are to preserve a consistent time series, we need to sacrifice (through amalgamation) a great deal of temporal detail. Conversely, if we wish to retain the maximum amount of spatial detail then we can only have short and broken time series" (Cliff and Haggett 1996, 332).

Researchers using GIS in a historical context have also argued for the need to understand the temporal as well as the spatial. Healey and Stamp argue that to understand regional economic development fully, the researcher must be able to disaggregate through both space and time as much as possible (Healey and Stamp 2000). This, they argue, involves looking at thousands of individual firms, preferably on a monthly basis over an extended period of time. Doing this requires being able to incorporate the rapidly changing locations of the firms and also developments in the transport network and changes in land ownership. MacDonald and Black researching the history of the book and print culture, come to broadly similar conclusions (MacDonald and Black 2000). They argue that a spatial and temporal framework is needed because to understand the development of print culture over time one must understand the complex relationships between such diverse activities as migration and other socio-economic variables (especially literacy rates), information about people employed in the book trade (location, occupation, and gender), the growth of libraries (location, type, and size), newspaper circulations, and so on, as they develop over time.

5.3 TIME IN GIS

Peuquet (1994) argues that a fully temporal GIS would be able to answer three types of queries:

1. Changes to an object such as 'has the object moved in the last two years?', 'where was the object two years ago?' or 'how has the object changed over the past five years?'
2. Changes in the object's spatial distribution such as 'what areas of agricultural land-use in 1/1/1980 had changed to urban by 31/12/1989?', 'did any land-use changes occur in this drainage basin between 1/1/1980 and 31/12/1989?', and 'what was the distribution of commercial land-use 15 years ago?'
3. Changes in the temporal relationships among multiple geographical phenomena such as 'which areas experienced a landslide within one week of a major storm event?', 'which areas lying within half a mile of the new bypass have changed from agricultural land-use since the bypass was completed?'

Unfortunately, the layer-based data model used by GIS does not allow easy handling of queries of this type and relatively little progress has been made in this direction. The basic problem relates to topology. GIS handles space efficiently by incorporating spatial topology (see Section 2). To also handle time efficiently it would need to have spatio-temporal topology. Although some suggestions have been put forward for doing this, mainly based on object-orientated technology (see, for example, Egenhofer and Golledge 1998; Wachowicz 1999) these have not yet been incorporated into GIS software.

5.4 METHODS OF HANDLING TIME IN HISTORICAL GIS

Although the temporal functionality included in most GIS software packages is usually very limited, there are many ways that time can be handled with a GIS. The best method to choose will depend on the nature of the individual researcher's data.

Spatial data

Attribute data

ID	Start Date	End date	Name	Output
1	1/1/1870	31/12/1870	Smiths	870
1	1/1/1871	31/12/1871	Smiths	930
1	1/1/1872	31/12/1872	Smiths	990
2	1/1/1870	31/12/1870	Jones	405
2	1/1/1871	16/5/1871	Jones	115
3	1/1/1870	31/12/1870	Frasers	610
3	1/1/1871	31/12/1871	Frasers	540
3	1/1/1872	30/6/1872	Bloggs	205
3	1/7/1872	31/12/1872	Bloggs	365

Figure 17: Time as an attribute. This example shows the history of three firms over time. Usually the data are in the form of annual output data, as happens consistently for firm 1. Where a firm opens or closes this can be represented using the attribute data: for example, firm 2 closes on 16 May 1871. Changes to the names of firms can also be handled in this way: for example firm 3 has its name changed from Frasers to Bloggs in 1872.

One simple way of handling time is to treat it as an attribute. Healey and Stamp do this in their study of regional economic growth in Pennsylvania (Healey and Stamp 2000). For both firms (represented as points) and railroads (represented as lines) the dates of their founding and closure are attached to the spatial features as part of the attribute data. In this way the development of the transport network and industrial development can be examined over time and the links between the two can be studied.

The simplest way to implement this is with a single row of attribute data attached to each spatial feature. Multiple rows can also be attached to each spatial feature with each row having a start and end date. This allows us to handle complex situations, for example, where the aim is to monitor a firm's economic statistics, such as output, profit, and employment, but where the name and ownership of the firm also changes over time. A simplified example of this is shown in Figure 17. Handling time in this way allows spatial features to be created and abolished and their attributes to change over time. The limitation of this approach is that the location of features cannot change.

Where the temporal nature of the data is more explicitly spatial, different layers can be used to represent the situation at different dates. This is termed the *key dates* approach and is particularly suitable where spatial data are taken from source maps from different dates. A good example of this approach is taken by Kennedy *et al.* 1999 in their atlas of the Great Irish famine. The atlas uses census data to show demographic changes resulting from the famine. At its core are layers representing the different administrative geographies used to publish the censuses of 1841, 1851, 1861 and 1871. These layers are linked to a wide variety of census data from these dates. This allows sequences of maps to be produced showing, for example, how the spatial distribution of housing conditions and use of the Irish language change over time.

While this approach is simple and effective, it is only suitable for a limited number of dates or where change occurs at clearly defined times between periods of relative stability. More complex situations are more problematic. If, for example, a researcher wanted to create a

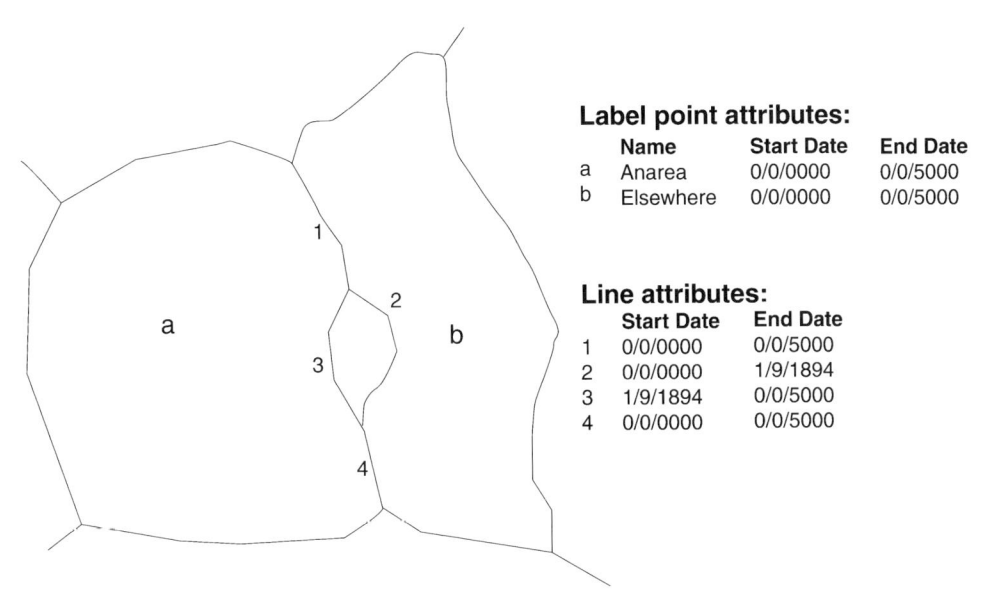

Figure 18: Storing changing administrative boundaries: the Great Britain Historical GIS. A master layer consists of label points (representing administrative units), and lines (representing their boundaries). The attributes for both sets of features includes date stamps. Features in existence when a type of unit was formed are date stamped 0/0/0000, while those in existence when they were abolished are date-stamped 0/0/5000. There is no topology on a master layer. This example shows how a boundary change between Anarea and Elsewhere on 1 September 1894 can be handled. Relevant features are selected from the master layer for a user-specified date and topology is then constructed. Source: Gregory and Southall 2000, 327.

Label point attributes:

	Name	Start Date	End Date
a	Anarea	0/0/0000	0/0/5000
b	Elsewhere	0/0/0000	0/0/5000

Line attributes:

	Start Date	End Date
1	0/0/0000	0/0/5000
2	0/0/0000	1/9/1894
3	1/9/1894	0/0/5000
4	0/0/0000	0/0/5000

database of changing administrative boundaries for an entire country, the key dates solution would be to digitise the boundaries for every date at which maps are available. There are two problems with this: first of all, where boundaries do not change the same line has to be digitised multiple times. This results in redundant effort and will inevitably lead to problems with sliver polygons (see Section 4). Secondly, it is unlikely to be possible to digitise the boundaries for every possible date, and while linking attribute data to spatial data for a nearby date may provide a good approximation of the actual boundaries, there will be some error introduced as a result. This can range from an incorrect representation of the administrative unit concerned, to either polygons with no attribute data or attribute data with no polygons.

These limitations have led researchers in various countries to attempt to build systems that are continuous records of boundary change. This allows the researcher to extract the boundaries for the appropriate date and link them to any suitable attribute data. Two distinct approaches can be identified to doing this: the *date-stamping* approach used by the Great Britain Historical GIS (Gregory and Southall 1998; 2000), and the *space-time composite* approach which was proposed as a theoretical structure by Langran (1992). This approach has been used by the Swedish National Topographic Database (Kristiansson 2000), the Belgian Quantitative

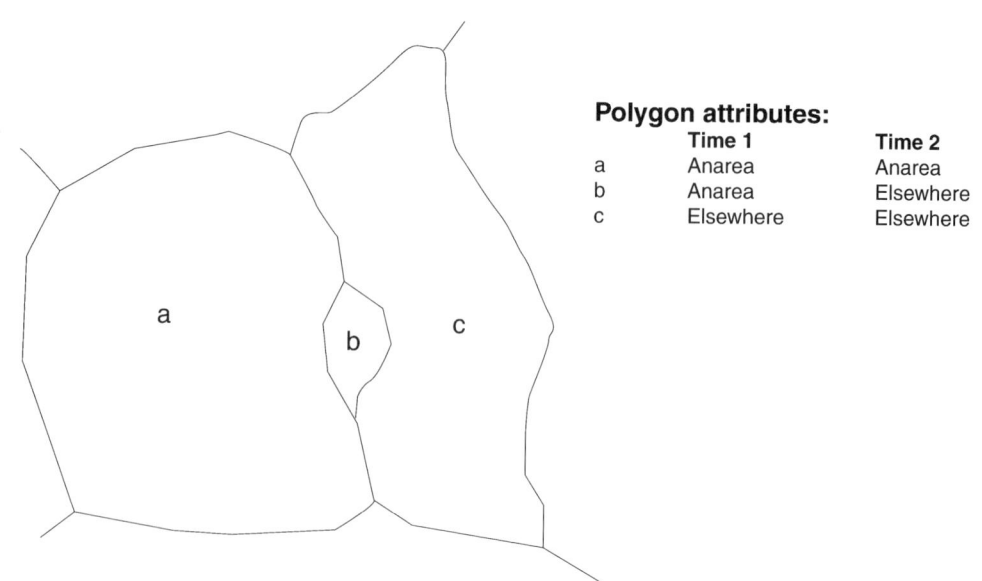

Figure 19: Storing changing administrative boundaries: a space-time composite. This diagram shows the same change as Figure 18. Using a space-time composite the Least Common Geometry is a polygon layer. The attribute data are split into time periods and each polygon is assigned to an administrative unit for each time period. The user specifies a time period and a dissolve operation (see Chapter 4) is used to aggregate the polygons to form the administrative units in existence during the specified period.

Polygon attributes:

	Time 1	Time 2
a	Anarea	Anarea
b	Anarea	Elsewhere
c	Elsewhere	Elsewhere

Databank (Vanhaute 1994) and is proposed by Ott and Swiaczny for the Palatinate area of Germany (Ott and Swiaczny 2001).

The date-stamping approach handles time as an attribute in a manner similar to that described above; however, it is complicated by the need for polygon topology. Gregory and Southall (2000) cope with this by storing all their spatial data in what they term *master coverages* (i.e. master layers). These are layers that have both label points, representing administrative units, and lines, representing their boundaries. There is, however, no topology to link the two at this stage. In this structure, boundary changes can be handled in the manner shown in Figure 18, showing a transfer of territory from one unit to another. More complex changes, such as where an entire unit is created or abolished, can be handled using the label point attributes. Within this structure, a user can specify a date and custom-written software extracts the appropriate label points and lines and creates topology to form a polygon layer for that date.

The space-time composite approach creates administrative units through aggregating smaller polygons, by storing the unit that each polygon lies in at each date as attribute data. These smaller polygons are referred to as the Least Common Geometry (LCG). This can consist of low-level administrative units that are known to be stable over time, as in the Swedish system that uses parishes to create districts, municipalities, and counties. Where no such units are available, it can consist of polygons created as a result of boundary changes, as is proposed by the Palatinate system. In either case the basic structure is the same, as is shown in Figure 19. A dissolve operation is used to re-aggregate the polygons in the LCG to form the units in existence at the required time.

5.5 CONCLUSIONS

Historical research has frequently been hampered by its inability to manage data effectively through attribute, space and time simultaneously. This has meant that, traditionally, researchers have had to simplify at least one of these three components in order to perform their analyses. GIS opens up a wide new potential for managing data through all three components without having to resort to simplifications. Although time is currently poorly integrated into GIS software, there is still real potential for using GIS to manage complex spatio-temporal datasets, as is illustrated by the case studies provided. As will be discussed in later Sections, this opens up the potential for exploring, analysing and visualising complex spatio-temporal change in a more sophisticated manner than has previously been possible. This should ultimately provide a more detailed and less simplistic understanding of the processes that drive these changes.

Section 6: Visualisation from GIS

6.1 INTRODUCTION

The map is an absolutely integral part of GIS. As a result the use of maps to explore and present GIS data, and the results of GIS-based analysis, is an absolutely core part of the GIS toolbox. It is important to remember that visualisation in GIS is about more than simply mapping: graphs, tables, and all the more conventional methods of displaying data are also valid and useful tools. This Section will, however, focus on mapping. In GIS mapping is used to describe and suggest explanations for the spatial impact of the subject of interest. This re-opens an enormous amount of potential to explore and explain the geography of a wide range of historical themes, an approach that has arguably been over-looked in the past. This Section aims to show how, through a combination of mapping and thinking spatially, new understanding can be developed.

6.2 MAPPING AND CARTOGRAPHY IN HISTORICAL RESEARCH

The map is a powerful way of presenting the information held within spatially referenced data to an audience. Cartography is an academic discipline in its own right with a long history. It is both a science and an art. From a scientific perspective, its role is to present features on the earth's surface to an audience in an accurate and objective manner. From an artistic perspective, its role is to present this information in a way that is both communicative and pleasing to the eye. These two roles are sometimes contradictory, and it requires skilled use of cartographic principles to balance these two objectives.

Most GIS software packages make it easy to produce basic maps as it is a core part of their functionality. This means that almost as soon as the data are in a GIS format, researchers are able to explore such data through maps. The maps can be refined and re-drawn multiple times as part of the research process, giving the researcher the ability to gain a thorough understanding of the spatial patterns the data contain. At the end of the research process, production of maps for publication either on paper, or more recently electronically on the Internet or CD-ROMs, becomes a relatively simple process.

This means that historians wanting to use GIS need to learn the basics of cartography, so that the maps they create and interpret lead to improved understanding rather than misleading or causing confusion. In this Section it is only possible to explain briefly a few basic rules about how good-quality maps can be produced. Many good guides to cartography are available, and the bibliography lists some of them.

A map can be regarded as a simplified abstraction of the world which presents complex information about one or more phenomena in an understandable manner, and which is also a valid picture of the underlying data. To do this effectively it is important to follow a number of general rules:

1   The map should contain as much detail as is necessary but not so much that the pattern becomes obscured, cluttered or over-complicated.

2   A map should stand alone and be understandable without referring to the accompanying text. To this end it needs a title, a legend and representation of scale. The legend should explain all symbols and shading used on the map.

3   The method of symbolisation used should be appropriate for the data being represented. GIS usually simplifies this. Points are usually represented by point symbols, polygons are represented using *choropleth maps*, and continuous surfaces are often represented using *isolines* (for example, contours to show relief).

4   The symbols and shading used should be as self-explanatory as possible to minimise the amount of times that the user needs to refer to the legend. For example, water features should be coloured blue.

5   If a shading scheme is used to represent a hierarchy, the features at the bottom of the hierarchy should be shaded in the lightest colours, those at the top in the heaviest, and there should be a clear and self-apparent progression up the hierarchy.

6   Where a continuous variable, such as the unemployment rate, is sub-divided into discrete classes, care should be taken in both the choice of the number of classes and in how they are defined. In general, for grey-scale maps no more than four or five classes should be used. Where colour is available this may be increased if necessary but never to more than ten. The intervals should not be arbitrarily chosen but should rely on some characteristics of the data. Examples include putting equal numbers of observations in each class; using equal intervals (for example if the range of a variable is from 0 to 20 and four classes are required the breaks would fall at 5, 10, and 15); using the mean and a standard deviation either side of the mean, and so on. The choice depends on the frequency distribution of the data, with heavily skewed datasets being among the most difficult to represent. Evans provides a detailed discussion of this (Evans 1977).

6.3 DEVELOPING UNDERSTANDING FROM BASIC MAPPING THROUGH GIS

Although it is rarely efficient to create GIS data simply to produce a map, once data are in GIS formats their potential use for mapping purposes becomes enormous. In particular, once it is easy to map a dataset then exploring the geography within the data becomes easy. A simple example of this is provided by Spence (2000a). He has a single layer of spatial data showing a simplification of the administrative geography of London in the 1690s. Linking this to taxation data allows him to explore the social geography of London at the time. This is done by following themes such as household rents, business rents, household densities, distribution of households by gender, location of widows, and so on. Producing maps also allows possible explanations to be developed: for example, Spence finds high concentrations of widows in the City and to the east of London. He speculates that there may be different explanations for this in these two areas. In the wealthy City it may reflect the dominance of males as property holders in this area, even after the men concerned had died. In the less affluent east, however, he speculates that this may reflect the dangers of the types of employment available to men from this area.

Mooney follows a similar approach (Mooney 2000). He links spatial data consisting of Registration Districts in London in the late 19th century to data on admissions to various hospitals. He uses this information first to describe the patterns of admissions to the London hospitals, and then to attempt to describe the spatial patterns of various diseases in the metropolis at that date. In the same volume, Galloway investigates London's national importance as an economic centre in the medieval period (Galloway 2000). This is done by producing dot maps of a variety of debt statistics from around 1400 to illustrate the economic interactions between London and the rest of England.

These three essays are all good examples of how basic mapping and thinking spatially can provide new insights into a discipline without the need for complex analyses.

6.4 PRODUCING ATLASES FROM GIS

Extending the above approach allows atlases to be produced using a small amount of spatial data coupled to a wide variety of attribute data. A good example of this is Woods and Shelton (1997). They use a single generalised layer of 19th century registration district boundaries in England and Wales to produce an atlas of mortality in Victorian times. By looking at the spatial pattern they are able to provide new insights into phenomena that have a pronounced geographical pattern. This atlas shows the importance of spatial detail and the power of maps to present it. There were over 600 registration districts in England and Wales in the Victorian era, and the maps shade all of them. This allows the authors not only to comment on the general patterns, but also on specific details and exceptions. When looking at infant mortality in the 1860s, for example, the authors note that the maps show a general pattern of high rates in urban districts and lower rates in rural ones. Looking in more detail reveals exceptions to this; for example, there were high rates found in rural areas around the Wash, in Lincolnshire and East Yorkshire, and in Cornwall. This kind of detail is readily apparent from maps but is difficult to spot in other ways.

The atlas of the Great Irish Famine by Kennedy *et al.* follows a similar approach, but rather than using a single layer of spatial data they use different layers for different dates (Kennedy *et al.* 1999; see also Section 5). By mapping changing housing conditions they show that on the eve of the famine the lowest quality housing, mud huts and similar constructions, were concentrated in the west of Ireland where they could form up to 50% of the housing stock. In the east, housing of this quality was less common. The poor who lived in this housing were the most seriously affected by the famine, so as the famine progressed this class of housing all but disappeared in the east, while in the west its importance also declined but to a lesser extent.

In the above examples the atlases produced are cartographically simple products based around choropleth mapping. GIS can also contribute to atlases that are more sophisticated cartographically. Here, rather than linking a large number of attribute datasets to a limited number of spatial layers, the GIS becomes a database of the spatial features that will be used in the atlas and allows them to be combined to produce highly sophisticated cartographic products significantly more cheaply than through traditional methods. A good example of this is volume II of the *Historical Atlas of Canada* (Louis 1993), which was produced using *ArcInfo*. Pitternick explains the advantages that using this software gave to this volume, compared with volumes I and III which used more traditional methods (Pitternick 1993).

6.5 ELECTRONIC VISUALISATION FROM GIS

With a paper atlas, the authors produce one or more maps, some text, and perhaps some diagrams, and use these to tell a story. Electronic media and GIS allow the author to present the user with the spatial and attribute data and allow the user to produce maps and diagrams themselves. This allows them to explore the data and, perhaps, tell their own story or investigate the places and themes they are most interested in. Effectively this involves giving the user prepared spatial and attribute data held in a software package that has the basic mapping and querying facilities of GIS software but is easy to use and, perhaps, steers the user down certain routes. A good example of this approach was produced by Gatley and Ell (2000). They produced a system on CD-ROM that contains a variety of census, poor law and vital registration data from 1801 to 1871. The statistics provide the attribute data, with polygons representing the administrative units providing the spatial data. The package allows the user to query the data to produce maps, graphs, pie-charts, and other diagrams. If the user creates maps, the shading and class intervals can be changed and individual polygons can be queried. In the United States, the Great American History Machine (Miller and Modell 1988) was intended to be a similar system. This held census data and statistics from the presidential elections from 1840 to 1970 at county-level, giving over 3,000 units. Users were free to use these data to produce choropleth maps of their own to explore the data. Unfortunately, this system was never properly published commercially.

6.6 OTHER FORMS OF MAPPING

Although choropleth maps are a highly effective way of communicating information from polygon-based data they do have certain drawbacks. A major problem with many administrative geographies is that the largest administrative units tend to be sparsely populated rural units, while the smallest tend to be densely populated urban areas. This means that the map can distort the pattern as it emphasises the areas in which few people live while almost obscuring the areas containing most of the population. One way around this is to distort the map pattern to make the polygon sizes proportional to the size of their population rather than their surface area. A map of this sort is called an *area cartogram*. An automated method of doing this while retaining perfect connectivity between each polygon and its neighbours has not yet been devised. Dorling presents a method where each polygon is converted into a circle whose size is proportional to its population (or any other variable) (Dorling 1994; 1996). The position of the circles is then moved to prevent circle overlap while attempting to keep a circle as close as possible to its neighbours. Dorling (1995) presents an atlas of modern Britain based almost

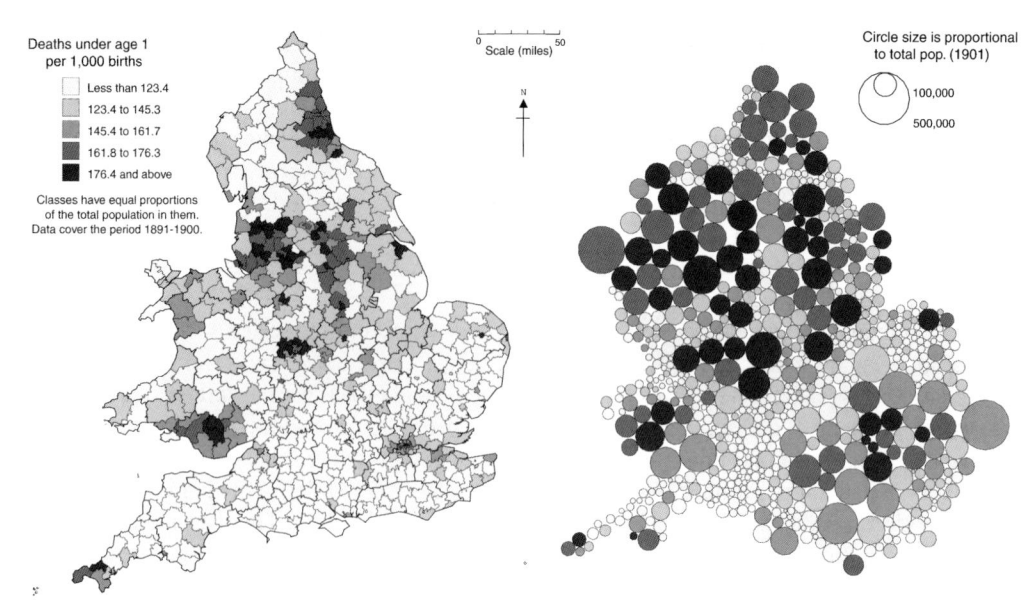

Figure 20: Choropleth and cartogram representations of infant mortality in England and Wales, 1890s. Both maps show the same pattern. The choropleth makes it appear that the highest rates of infant mortality are rare. The cartogram shows that they are in fact common in urban areas where the majority of the population live. In fact, the five classes contain approximately equal numbers of people. Source: Gregory et al. 2000, 140.

entirely on cartograms but is otherwise very similar to atlases such as Kennedy *et al.* (1999), Spence (2000b) and Woods and Shelton (1997).

Cartograms can be criticised as they are an unfamiliar representation of space and it can be difficult to establish exactly where a place is on the map. Gregory *et al.* use choropleths and cartograms together in their study of changing patterns of poverty (Gregory *et al.* 2000). Figure 20 gives an example of this showing registration district-level infant mortality in the 1890s. From the choropleth map it appears that the highest rates of infant mortality are relatively rare. The cartogram gives a very different impression by showing that because high rates of infant mortality were largely an urban phenomenon, a large proportion of the population were in fact affected by these rates. In actual fact, the shading scheme used by these maps puts approximately the same number of people into each of the five classes.

6.7 MOVING AND INTERACTIVE IMAGERY

Animation is an area that has much to offer historians. Traditional paper-based cartography is poor at presenting change over time, while animations clearly have the potential to do so. If one aim of GIS is to be able to explore and present data with the maximum amount of both spatial and temporal detail possible, then this may seem like an ideal way of approaching the subject.

There are a large number of modern file formats that allow the researcher to create animations, from *animated GIFs*, where a number of GIF images are stitched together sequentially, through to more complex video formats such as *AVI* and *MPEG*. While the technical issues associated with producing animations are well developed, the cartographic issues are still in their infancy. As a result, researchers working in this field need to be careful to remember that the aim of producing animations is the same as with any other type of mapping: to produce a clear and understandable abstraction of the data and the patterns held within them. It is easy to become seduced by fancy graphics that convey little of real value, and the cartography of animation is, as yet, poorly understood.

Moving imagery can be used to do more than simply present time series. It can also be used to change view points or even to fly-through virtual landscapes. A good example of this is provided by Harris (2000). He is researching an ancient burial mound and sacred space in Moundsville, West Virginia, and wanted to re-create the landscape that would have existed around the mound. By combining historical and archaeological evidence on the ancient landscape with modern data on relief he is able to create a digital terrain model of the landscape. This is draped with a representation of the possible land cover of the time. The user is then able to zoom in and out of the landscape and fly-through it to view it from any angle or perspective that they are interested in. As with animations, this is an interesting research area but efforts must be put into conveying information to the user and not simply producing fancy graphics with gimmicky file formats.

There has been increasing interest over time in mapping on the Internet. Maps and images such as those produced by Ray are indeed available over the web (Ray 2001). While the web is a good medium for raster images and a variety of animation and multimedia formats, as yet putting vector graphics on the web is still fraught with problems, despite vendor claims to the contrary. This is partly a problem of reliability and partly a problem of performance. Java offers one way of putting vector images onto the web (see Southall *et al*. 2001); however, these require low-level programming skills. Once it is technically easier to put vector graphics on the web there will still be many cartographic issues to resolve.

6.8 CONCLUSIONS

Mapping and visualising data and exploring them spatially allow new insights into the patterns contained within those data. This is an area that has traditionally been under-exploited by historians and it is hoped that the increasing use of GIS will lead to an increased awareness in the importance of space and geographical patterns among those conducting historical research. The ease of mapping and visualising data within GIS is leading to a change in the role of the map and in map authorship. Already, GIS has moved the map from an end product of a piece of research to an integral part of the process. Increasingly, it is leading to a changing role of authorship. Traditionally research papers and atlases present a map prepared by the researcher together with some text in which the researcher explains what he or she believes the map shows. The advent of electronic technology is changing this. CD-based products such as SECOS and Internet-based resources such as Ray (2001) mean that the researcher is no longer attempting to tell the whole story. Instead the researcher makes the resources available to users (or readers) who interrogate the data themselves and use them to tell their own story.

The final word in this Section, however, must be a word of warning. Maps distort and maps lie. Whether you are producing maps yourself or interpreting information from other people's maps, it is important to have a grasp of the basics of cartography to be able to differentiate between the distortions that the map contains (deliberately or accidentally) and the real story that the data have to tell.

Section 7: Spatial Analysis of Statistical Data in GIS

7.1 INTRODUCTION

Visualisation helps us to understand spatial patterns but often further investigation is needed to describe or explain spatial patterns. When using quantitative data this frequently means that researchers will want to perform some form of statistical analysis. When using GIS, an analysis should use not just conventional statistical techniques, which only focus on the attribute data, but should also incorporate the spatial component of the data. Bringing the two together is known as *spatial analysis*, or *geographical data analysis* (GDA).

Caution is required when using GIS-based analysis of spatial data, and lessons can be learned from the experience of GIS research in human geography. Early enthusiasm for a GIS-based approach led protagonists to claim that GIS provided a cohesive and scientific framework that could re-unite geography as a discipline. This was presented within a highly quantitative framework (see, for example, Openshaw 1991). Not surprisingly, this led to a backlash that particularly focused on the overtly quantitative and scientific approaches that GIS research was taking at this time. This, it was argued, marked a return to "the very worst sort of positivism" (Taylor 1990, 211). It was also argued that as an academic sub-discipline, GIS lacked a strong epistemology and any treatment of ethical, economic or political issues (see, in particular, Pickles 1995).

As time has gone on, some of the more extreme claims about what GIS may achieve have been moderated, while some of the criticisms of it are being addressed. In particular, it has become clear that GIS has more to offer than simply being a quantitative, positivist, number-crunching operation and, as will be discussed in Section 8, it has a potentially significant role to play in qualitative research as well as quantitative. In addition, through Geographic Information Science (GISc, see Section 1), the role of quantitative GIS-based work as an academic sub-discipline is being increasingly well defined.

While there has been much enthusiasm for including statistical functionality as part of the GIS toolbox among researchers, this has not been mirrored among the GIS software vendors. Many software packages claim to have analytic functionality but this usually refers to the inclusion of overlay, buffering and the other forms of spatial manipulation described in Section 4, rather than the more statistical techniques dealt with in this Section. This has advantages as well as disadvantages. In particular, it encourages researchers to determine for themselves what types of techniques are relevant to their data, and to devise methods of performing the analysis either within the GIS software or outside it. Effectively this means that at present we are forced to think about what techniques are appropriate to our data rather than simply being spoon-fed with vendor-provided 'solutions' that may or may not be appropriate. While this may lead to cumbersome computing, it is hoped that it should lead to more appropriate analyses.

7.2 WHAT MAKES SPATIALLY REFERENCED DATA SPECIAL?

Spatially referenced data have special characteristics that offer both advantages and disadvantages to the researcher wanting to perform statistical analysis. The advantages are basically that they allow us to ask questions such as 'where does this occur?', 'how does this pattern vary across the study area?', 'how does an event at this location affect surrounding locations?', and 'do areas with high rates of one variable also have high rates of another?'. Conventional statistical techniques such as *correlation* and *regression* tend to produce a summary statistic that quantifies the strength of a relationship within a dataset or between two of more sets of variables, for example, a correlation coefficient of 0.8 between X and Y. This approach is termed *global* or *whole-map analysis* and is undesirable from a GIS perspective because it ignores the impact of space by simply providing a summary of the average relationship over the whole study area. In reality any relationship is likely to vary over space and when performing analysis on spatially referenced data we should attempt to use techniques that allow us to examine not simply what the average relationship is over the whole map, but how the relationship varies across the map. This is termed *local analysis* (Fotheringham 1997). Some examples of how this can be done are given in section 7.3. For now it is sufficient to make the point that by using techniques that explicitly incorporate the spatial component of the data we are able to use the data to develop a more sophisticated understanding. Ignoring the effects of space, as many traditional techniques do, means that we are limiting the knowledge that can be gained from the data.

There are four major disadvantages that make spatially referenced data special. These are: data quality and *error propagation*; *spatial autocorrelation*; the *modifiable areal unit problem* (MAUP); and *ecological fallacy*. Many of the issues regarding data quality with spatially referenced data were discussed in Section 4. Here it is sufficient to say that there is always error, inaccuracy, and uncertainty in the spatial component of data. When data are combined through overlay or similar operations the error and uncertainty become cumulative, a process known as error propagation. GIS software packages do not handle this well, as their data model only allows a single concisely defined representation for each spatial feature, so the possible impact of error must always be considered.

Many statistical techniques make the assumption that the observations used in the study are independent pieces of evidence. This is usually referred to as being 'independently random'. If we are studying the locations of the data to try to find a geographical pattern then we are working on the principle that the locations are being influenced by some underlying cause that varies over space. The data are referred to as being spatially autocorrelated, and this invalidates the assumption of independent randomness. Spatial autocorrelation is similar to temporal autocorrelation but is more complicated, as it operates in many directions simultaneously. The degree of spatial autocorrelation in a dataset may be quantified, as is described later in this Section.

Many socio-economic datasets, for example the census, are published as totals for administrative units. The boundaries of these units are, to all intents and purposes, arbitrary and random, defined as a result of politics and inertia rather than because they say anything meaningful about the population that they are sub-dividing. The question this raises is: if the arrangement of the administrative units were changed, would the results of any analyses based on them also change? Openshaw and Taylor (1979) vividly demonstrated that it could. They compared data on the percentage of the population voting Republican in the 1968 Congressional election with the percentage of the population aged over 60 for 99 counties in Iowa. By aggregating these data to six regions using different arrangements they could produce correlations from -0.99 to 0.99 and almost any result in between. In other words, the results of the analysis were totally dependent on the way in which the data were aggregated. Similar phenomena have also been demonstrated using regression analysis (Fotheringham and Wong 1991), and they occur as a result of two effects combining. The first, the scale effect, simply means that as data are aggregated they become increasingly averaged or smoothed as extreme values are merged with more normal areas. The second is connected to the actual arrangements of the boundaries, which can lead to results being 'gerrymandered' in a similar way to election results. Taken together these effects are referred to as the modifiable areal unit problem (MAUP).

At one level the MAUP is highly worrying, as it means that the results of any analysis using spatially aggregate data are highly suspect and could simply be the result of the administrative units used to analyse the data. Some statistical and GIS-based approaches to the MAUP have been suggested (see Openshaw and Rao 1995; Fotheringham *et al.* 2000), but these are either statistically or geographically complex and are not entirely satisfactory. A more pragmatic approach is to avoid additional aggregation by using the data in as close to their raw form as possible while interpreting the results of any analysis bearing in mind the possible impact that modifiable areal units may be having.

Ecological fallacy is closely related to the modifiable areal unit problem. It deals with the fact relationships found in aggregate data may not apply at the household or individual level. For example, if areas with high rates of unemployment also have high crime rates it is a mistake to think that a person who is a criminal is more likely to be unemployed than employed. This has been known since the 1950s, yet the increasing potential to analyse spatially aggregate data in GIS means that there is a temptation to ignore it. There have been attempts to find mathematical or statistical solutions to ecological fallacy but, as with the MAUP, these provide computationally and statistically complex ways into the problem without actually resolving it.

7.3 SPATIAL ANALYSIS TECHNIQUES

Having introduced the advantages and disadvantages associated with the statistical analysis of spatially referenced data, a few types of techniques that can be used are described here. It is not the intention to describe them in great detail as they are well described elsewhere (see the bibliography), but to give a flavour of the types of approaches that can be used.

The spatial data often determines what type of spatial analysis technique is most appropriate as different techniques must be used to take advantage of the characteristics of point, line and polygon datasets. It is sometimes sensible to analyse polygon data as points based on their centroids or points as polygons using Thiessen polygons.

Point pattern analysis is concerned with attempting to determine whether the distribution of points is random or whether it either clusters (positive spatial autocorrelation) or is evenly distributed (negative spatial autocorrelation) as shown in Figure 21. The easiest way of doing this is termed *quadrat analysis* whereby the study area is sub-divided into regular grid squares and the number of events in each square is counted. This is not particularly satisfactory, as the results tend to be heavily dependent on the size and arrangement of the grid squares. Better techniques focus on the distance between each point and its nearest neighbour or on kernel

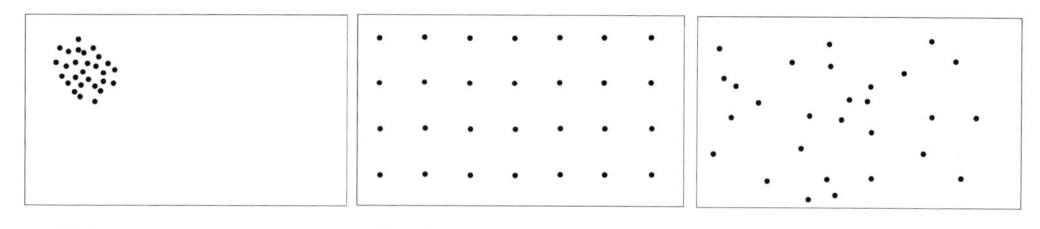

a. Positive   b. Negative   c. Random

Figure 21: Types of spatial autocorrelation in point patterns. With positive spatial autocorrelation all the points are clustered together, with negative spatial autocorrelation they are evenly distributed, while with random spatial autocorrelation there is no clear pattern.

estimations, which effectively produces a moving average of the density of points at each location on the map. Global summary statistics from these techniques will simply say that the data clusters, and is evenly distributed or is randomly distributed. It is better to use local techniques that allow summaries such as 'the data cluster in this location but are randomly distributed here'.

Pure point pattern analysis is simply concerned with analysing the spatial component of the data and ignores attribute. This is not possible with polygon data where the arrangement of the polygons will usually be arbitrary, such as with census data based on administrative units. For these it is necessary to devise *proximity measures* that allow us to quantify the influence that each polygon has on its neighbours. These can either be based on the distance between the centroids or based on some measure on the relationship between polygon boundaries. They may also be binary where polygon *i* either has an influence on polygon *j* or it does not, or there may be some degree of quantification on the strength of the relationship. Simple binary proximity measures include whether two polygons share a boundary or whether their centroids lie within a set distance of each other. More complex measures include ratios based on the length of shared boundary between two polygons compared to their total perimeters, or distance decay models whereby the influence one polygon has on another declines as the distance between the two centroids increases.

Once a proximity measure has been calculated, this may be used to analyse the polygon or point data set. A simple form of analysis is to test for spatial autocorrelation using techniques such as *Geary's coefficient* or *Moran's coefficient* that analyse proximity and attribute together. Traditionally these provide global summaries but techniques such as *Geary's G_i* allow us to perform local analysis of these questions (Fotheringham *et al*. 2000). More sophisticated forms of spatial analysis that combine the analysis of spatial and attribute data include *kriging* and *geographically weighted* regression (GWR). Kriging is a sophisticated interpolation technique that attempts to estimate a continuous trend surface from a set of known sample points (Isaaks and Srivastava 1989). GWR explicitly incorporates space into regression analysis. Rather than simply producing a single, global regression equation GWR produces an equation for each point or polygon in the dataset based on the location's relationship with its neighbours. This can be used to produce maps that show how relationships vary over space (Brunsdon *et al*. 1996).

Analysis of line data is often different from analysis of other forms of data, as it tends to concentrate on flows. This is termed *network analysis*. This can involve problems such as the way in which the shortest route between a set of points on the network is calculated, termed the *travelling salesman problem*. This can be made more sophisticated by classing lines on the network in different ways, for example, different types of road or railways may have different journey times or journey costs associated with them. Networks can also be used in *location-allocation models* that attempt to find the most efficient location on a network. An example of this might be calculating the most efficient location for an industrial complex based on certain assumptions about the rail network. These can then be compared with actual locations.

7.4 SPATIAL ANALYSIS IN HISTORICAL GIS

A variety of authors have made use of spatial analysis techniques to analyse historical data. Bartley and Campbell wanted to produce a multi-variate land-use classification for medieval England from a single source, the Inquisitions Post Mortem (IPM), which gave a detailed breakdown of landowners' estates on their death (Bartley and Campbell 1997). To give as comprehensive coverage of the country as possible they used 6,000 IPMs covering a 50-year period. These were given place names that allowed them to be allocated to points on the map. These points represent large areas with a variety of land-uses, and they wanted to create a more realistic and flexible representation of medieval land-use. To do this they reallocated the point data onto a raster surface, arguing that this offered a more valid representation of continuous data than either points or polygons. This was done using a kernel method that calculated the value of each pixel based on all eligible IPM points within 250 square miles of the cell, with nearer IPMs being given more weight than further ones. They then used cluster analysis to allocate each cell to one of six land-use classes. To take into account the uncertainty in their model, the technique they used allowed cells that were hard to classify to be given second choice alternatives. Through this imaginative and sophisticated use of spatial statistics they were able to create a detailed classification of pre-Black Death England that in turn will allow other studies of localities to be interpreted.

Bartley and Campbell use sophisticated handling of spatial data and statistical techniques to create a complex surface from a point layer. Cliff and Haggett use basic classifications, overlay and exploration through bar charts and basic statistics to integrate data from a variety of sources to explore the cholera epidemic in London in 1849 (Cliff and Haggett 1996). They believed that there were two main factors affecting the variation in severity of the epidemic in different parts of London: drainage, and the source of the water supply. To model poor drainage they created a polygon layer that distinguished high and low risk areas based on height above the level of the River Thames. London's water supply came either from reservoirs or wells, or directly from the Thames itself. Areas supplied by water extracted directly from the Thames were believed to carry a higher risk factor than those supplied from wells and reservoirs. Sub-dividing London in this way provides a second polygon layer. Overlaying these two layers sub-divides London into four types of area: those at risk as a result of both poor drainage and a polluted water supply, those at risk only as a result of poor drainage, those at risk only as a result of a polluted water supply, and those with neither risk factor. They then calculated the death rate from cholera in each type of area. Even simply graphing the data shows a clear pattern: areas with both risk factors had a death rate that was nearly twice as high as the metropolitan average. Those with defective drainage alone were also above average, those with polluted water were around average, and those with neither factor were well below. Statistical analysis using analysis of variance (ANOVA) confirms this pattern.

Gregory performs an analysis of census and vital registration data to show how manipulating the spatial and temporal component of the data can increase information about attribute (Gregory 2000). He does this by working on net migration over a 50-year period from 1881 to 1931. The census provides population figures split into five-year age bands and by sex. The *Registrar General's Decennial Supplements* provides the number of deaths sub-divided by age and by sex. In theory this gives enough information to calculate the net migration rate in ten-year age bands and by sex. Subtracting the number of, for example, women aged 5 to 14 at the start of a decade from the number aged 15 to 24 at its end and then subtracting the number of deaths in this cohort should give the net migration rate. Boundary changes make this calculation highly error prone, as any population change caused by a boundary change will appear to be net migration. By *areal interpolation* of all the data onto a single set of districts, age and sex, specific net migration rates can be calculated at district-level over the long term. This analysis, therefore, takes very basic demographic data and uses it to provide a long time-series of new data that have far more spatial and attribute detail than it was possible to create from manual methods.

Martin compares the 1981 and 1991 censuses (Martin 1996a). He interpolated data from the two censuses he was interested in onto a raster grid consisting of square pixels with 200m sides. In this way he was able to compare data between the two censuses with far more spatial detail than Gregory, but was unable to include earlier censuses as they did not provide sufficient spatial detail to allow accurate interpolation.

7.5 CONCLUSIONS

Although a significant amount of progress has been made in determining the research agenda for quantitative analysis of statistical data, there is still some way to go before this functionality becomes a standard part of the GIS toolkit, and even further before it becomes a common part of historical analyses. There is a large amount of potential for the statistical analysis of spatially referenced data, but the limitations of the data must be remembered. This means that researchers must select techniques that do not rely on the assumption of independent randomness, and must bear in mind the limitations of their data, in particular: its quality; potential problems caused by error propagation; the possible impact of modifiable areal units; and the potential for incorrect inference due to ecological fallacy.

As the use of GIS among historians becomes increasingly sophisticated, it is to be expected that increasing use will be made of spatial statistical techniques. The fact that these are not well integrated into GIS software as yet provides opportunities for researchers to devise the techniques that are most suitable for their own data, although these must be devised within the limitations of the source data. The impact of the spatial nature of the data must be borne in mind here and, in particular, unnecessary aggregation should be avoided not only through space, but also through attribute and time. Such techniques should not just produce a single summary statistic for the entire study area but should show how the impact of the phenomenon varies across the map. The results of the technique may be map-based rather than using traditional numeric summaries.

Section 8: Qualitative Data in GIS

8.1 INTRODUCTION

Section 7 described some well-developed quantitative techniques that could be used by a historian from a social science tradition. The use of qualitative data of the kind more likely to be more relevant to a researcher from the humanities end of the discipline is less well developed. The basic principles are the same: the GIS imposes a structure where data have spatial and attribute components and are organised in a series of layers, with each layer representing a different theme. This structure allows the user to explore the geographical relationships within the data. For historians working in social science this is likely to involve thematic mapping and analysis of statistical data in ways similar to those that have been well explored in other disciplines. For humanities scholars the data used are likely to be more complicated, comprising textual descriptions, images, photographs, maps, sound and video. Although these are different from statistical data, if they can be geo-referenced using points, lines, polygons or pixels, then GIS can be used to explore them. Mapping will again be an important part of this but so will querying the data to investigate the spatial relationships within and between layers of data and perhaps the calculation of summary statistics. Because far less research has been done in this area, this Section is significantly shorter than some others. This reflects the fact that historians working with qualitative sources are defining the research agenda as they go rather than applying techniques that originated elsewhere.

8.2 TYPES OF QUALITATIVE DATA IN GIS

Many qualitative projects follow a data-led approach. Often a large amount of disparate information about a place or places is gathered and integrated using the locational characteristics of the data. By assembling the data in this way it becomes possible to explore them to form new insights. Qualitative information is often point-based, but can also involve lines or polygons. Examples of the types of qualitative attribute data that may be used in GIS include information on the presence or absence of features such as roads and buildings, documents concerning events that occurred at a place, pictures or photographs relating to places, or audio or video that refer to a specific location. Textual descriptions of places from historical manuscripts can also provide data for a qualitative GIS. Historical maps or plans could also be part of the qualitative attribute data, particularly if they have relevance to an area but cannot be geo-referenced in the manner of modern-style maps. Rather than store all the information about a place in a single GIS, the GIS attribute data could consist of hyperlinks to websites that contain information about specific places or themes. All that is required to turn these into GIS data is a coordinate-based location.

Traditionally, the locational component of qualitative information is less likely to fit the precisely defined locations insisted on by GIS software. In some ways, this is less of a problem than it can be for quantitative data because the types of analyses that are performed are less likely to be as demanding of the spatial component than, for example, areal interpolation operations using quantitative data. In spite of this, the accuracy of the locational data, particularly data derived from different sources, still needs to be carefully considered.

The use of qualitative sources to date often represents a more sophisticated use of the *metadata* (see Section 9) about an object. For example, the catalogue for a collection of scanned photographs is likely to have included some information about the place that the photographs refer to, usually place names. By converting these place names into spatial data, probably points, we have the ability to create a layer of scanned photographs. As with other GIS applications, this allows us both to find more information about the collection of photographs themselves, for instance, by comparing images of locations that are near to each other; and it also allows us to integrate the collection of photographs with other collections of photographs or perhaps with other types of information.

8.3 CASE STUDIES

A basic ability of GIS is to provide structure to a single layer of data. Core to this is the ability to query features based on their location. The *Survivors of the Shoah Visual History Foundation* <http://www.vhf.org/> has used this idea to provide a way of finding useful information from a large and complex database of video clips. The database consists of over 50,000 unedited video accounts of the stories of holocaust survivors recorded in 57 countries in 32 languages. A fundamental problem with this database is how to allow users to find the information that they want in an efficient and flexible manner. The database uses two approaches: key-word searching and graphical user interfaces. The graphical interfaces have the advantage of being easy and intuitive to use. Geography has an important role in this as it allows the user to explore the database through map-based interfaces in order to access information about specific places. The GIS component is, therefore, a 'hub technology' (Lang 1995, 44) around which much of the database is structured.

GIS can also be used to integrate data from different sources through the use of multiple layers. The Perseus Project <http://perseus.csad.ox.ac.uk/> is a good example of this (Smith *et al.* 2000). It attempts to integrate various electronic libraries and archives from around the world and again uses locational information as a key component of this. One of the libraries they were concerned with was Edwin C. Bolles' collection of the history of London <http://perseus.csad.ox.ac.uk/cgi-bin/perscoll?collection=Bolles>, compiled in the late 19th century. It includes: printed sources, some of which are unique; folio descriptions of the city from limited print runs; contemporary 19th century maps; and illustrations and prints from the 17th to the 19th centuries. Again, using GIS as a hub technology allows the user to access these data efficiently and flexibly based on location. In the case of the Bolles' collection, it allows data from different sources to be brought together and explored. This involves having digital maps from different dates linked together, having place names identified in source texts with hyperlinks to other references to the same place, and having coordinate-based locations associated with images.

In the two projects described above the GIS is used primarily as an archival tool. The following two examples show it being used more directly as an exploratory analytic tool. One example focuses on the Salem witch-trials in Massachusetts in 1692. Ray describes an archive that includes complete transcriptions of contemporary court documents, transcriptions of rare books written about the trials, contemporary maps of the village, historical maps relating to the trial, and information including transcriptions, scans of documents and catalogue information from a wide variety of archives in different places (Ray 2001). Access to all of these is available through a single website: Salem Witch Trials – Documentary Archive and Transcription Project <http://etext.virginia.edu/salem/witchcraft/home.html>. The GIS allows the spatial aspects of the trial to be queried and understood. In particular, it allows the 300 people mentioned in the court records to be put into their actual households to provide a better understanding of the property disputes that many historians believe to have underlain the accusations. This includes information on the age and gender of the people concerned, the frequency of the accusations made by or against them, their family relationships, and the relative wealth of the accused and their accusers. Queriable maps and animations are available to allow the user to get a better understanding of the geographic nature of these relationships.

The Valley of the Shadow Project follows a similar approach <http://jefferson.village.virginia.edu/vshadow2/contents.html>. Here the interest is to compare the experience of two communities two hundred miles apart and on different sides of the American Civil War. These communities are based in Franklin County, Pennsylvania and Augusta County, Virginia. The aim of the project is to track the lives of soldiers and civilians from these two communities during the war and beyond. Available sources include census records, tax records, soldiers' dossiers, letters and diaries. A major difficulty was how to recreate these communities at a level localised enough to provide a context to individual lives. GIS provides one method. In 1870 the 'Hotchkiss' map of Augusta County was produced. This map was localised enough to name over 2,000 individual dwellings, many of which were private residences. The map was scanned and layers digitised from it, including points such as dwellings, churches, schools, mines and mills; lines such as roads and rivers; and polygons such as electoral districts that were used to map census data. Individuals could then be allocated to their houses and other sources about the individuals linked to these. Doing this "allows us to locate people *within* the county and not simply treat then as undifferentiated residents *of* the county" (Ayers *et al.* <http://jefferson.village.virginia.edu/vshadow2/ecai/present1.html>).

8.4 CONCLUSIONS

The role of GIS with qualitative data has yet to be fully developed; however, early studies suggest that there is considerable potential here. This potential falls into two main areas. One area is as a catalogue tool that allows electronic archives to be queried and explored in a geographical manner. This is the approach followed by the Survivors of the Shoah and Perseus projects. The second area is that it allows the historian to research the spatial side of a problem in a more integrated and usable way than would traditionally have been possible. This is demonstrated by the Salem witch-trials and the Valley of the Shadow projects' use of GIS to re-create local communities in order to gain a better understanding of events in the past. The use of GIS in qualitative research is not yet well developed, however, but the examples given here show that its power as an integrating and exploration technology and approach has considerable potential in this area, in addition to its more traditional role in humanities research.

Section 9: Preservation, Documentation and the Role of the History Data Service

9.1 INTRODUCTION

Section 3 makes it clear that creating or acquiring GIS datasets can be extremely costly in terms of both budgets and time. In later Sections methods by which existing GIS datasets can be taken were also described. These can take the form of integration with other datasets to produce new products (Section 4), creating new datasets from analyses (Sections 7 and 8) and creating visualisations that could potentially take the form of datasets (Section 6). In order that full use can be made of any dataset it is essential that it is adequately preserved and well documented.

The History Data Service (HDS) <http://hds.essex.ac.uk/> is funded by JISC <http://www.jisc.ac.uk/> to collect, manage, and encourage re-use of digital resources which result from or support historical research and teaching. The HDS is part of the UK Data Archive <http://www.data-archive.ac.uk/home/> at the University of Essex <http://www.essex.ac.uk/> and is the Arts and Humanities Data Service (AHDS) <http://www.ahds.ac.uk/> service provider for the historical disciplines, and one of its main roles is to preserve and disseminate historical datasets. The majority of the datasets within the HDS collection include a geographical component of some kind and can therefore be incorporated into a GIS, and it may be worth checking early on in the life of a GIS project to see whether the HDS hold any data that might be useful. The following section details how to obtain data from the HDS.

Similarly any datasets that are created, either by data capture or by modifying existing datasets, for use in a GIS can be deposited with the HDS to ensure their longevity and future use across changes in software and computing technologies. The third section explains the process for depositing data with the HDS.

The fourth section in this Section deals with the issues of *metadata* and documentation. These are integral and important elements of a dataset, GIS or otherwise, which enables someone who has not been involved in its creation to use it. At the most basic level these describe what the dataset contains, but they also describe the provenance of the data, the processes by which the data were captured, decisions made by the researchers when coding data and so on.

The final section of this Section provides details of how to obtain further information about these issues.

9.2 OBTAINING DATA FROM THE HISTORY DATA SERVICE

The HDS collection covers a wide range of historical topics with well over 500 separate data collections transcribed, scanned or compiled from historical sources. The collection includes databases, spreadsheets, electronic texts and scanned images, and covers the time period from the 8th century to the 20th century. Although the primary focus is on the UK, the collection includes a significant body of cross-national and international datasets, and it is particularly strong in 19th and 20th century economic and social history. Examples of topics covered include: 19th and 20th century statistics, manuscript census records, state finance data, demographic data, mortality data, community histories, electoral history, and economic indicators. Data can be located through the UK Data Archive Catalogue <http://www.data-archive.ac.uk/Search/searchStart.asp>, or through browse lists available at <http://hds.essex.ac.uk/studybrowse/>. Datasets are made available for use in research, learning and teaching.

To access data, users must first register with the HDS. This is a simple, online procedure which is done through the UK Data Archive, and which only needs to be done once. Once registered, users can order data from both the HDS and the UK Data Archive. For most academic users the only costs involved in ordering data are the price of the medium (CD-ROM or floppy disk), postage and insurance. Datasets are supplied in a variety of formats, including ASCII files, database files and spreadsheet files, and on a variety of media including CD-ROMs and floppy disks, as well as via FTP. For more information about obtaining data from the HDS, see Accessing Data <http://hds.essex.ac.uk/access.asp>.

9.3 DEPOSITING DATA WITH THE HISTORY DATA SERVICE

The role of the HDS is to collect, preserve, and promote the use of digital resources which result from or support historical research, learning and teaching. It is supported in this effort by the Arts and Humanities Research Board <http://www.ahrb.ac.uk/>, the Economic and Social Research Council <http://www.esrc.ac.uk/>, the Leverhulme Trust <http://www.leverhulme.org.uk/> and the Wellcome Trust's History of Medicine Programme <http://www.wellcome.ac.uk/>, all of which either require or recommend that their grant holders offer for deposit with the HDS any historical data that they may produce during the course of their funded project. The HDS also negotiates access to data held in the collections of other data centres and data archives.

There are a number of benefits to depositing data with the HDS.

Ensuring preservation

Firstly it ensures that the data are preserved. The time and resources invested in the creation of digital resources, and particularly with GIS datasets which can be very expensive to produce, can easily be placed in jeopardy because hardware and software become obsolete, and magnetic media degrade. Consequently it is in the interests of the data creator, potential users of the data, and research funding bodies, that the datasets generated are preserved in the long-term. This involves migrating the data to new hardware and software when the platforms on which the data were created become outdated, in order to maintain their usability into the future. This is a costly and complicated procedure, and is often beyond the resources of individual research projects that usually only run for a finite period. This is why many funding agencies make offering the data for deposit a condition of awarding grants to projects that involve data creation.

Providing access

Many historical GIS resources have significant and long-term value to the research and teaching community, and the time and resources invested in their creation can only be fully realised if they are systematically collected and disseminated. Access to such resources can facilitate communication within the research community, encourage the development of important new areas for research, and facilitate further research and debate of current research themes.

The HDS makes deposited data collections available for future re-use by distributing them in a range of formats and on a variety of media. Data collections deposited with the HDS are professionally catalogued, and information about them and any associated publications is made accessible through online catalogues, mailing lists and a variety of other publicity. This means that once the data have been created the maximum amount of use can be made of its potential. Again, given the high cost of creating GIS datasets, this is both important and desirable.

Data are deposited with the HDS with a non-exclusive licence for use in research and teaching. This means that all intellectual property rights and copyright are retained by the copyright holder(s), and that the depositor grants the HDS the necessary permissions to preserve and disseminate the data for use in research and teaching.

Professional recognition

Digital resources deposited with and preserved by the HDS gives professional recognition to the creator of the dataset. By collecting, evaluating, cataloguing and publicising data collections the HDS helps to provide tangible evidence of the scholarly effort involved in data creation. Data collections deposited with the HDS are widely publicised, for example, through workshops and online catalogues, and individual depositors gain professional recognition when their data collections are re-used in research and teaching and cited in subsequent publications.

For more information about depositing data with the HDS see its *Guidelines for Depositors* <http://hds.essex.ac.uk/depguide.asp>.

9.4 DOCUMENTING A GIS DATASET

The maintenance of comprehensive documentation detailing the data creation process and the steps taken involves a significant but profitable investment of time and resources. It is more effective if documentation is generated during rather than after a data creation project. Such an approach is likely to result in a better quality data collection as well as better quality documentation, because the maintenance of proper documentation demands consistency and attention to detail. The process of documenting a data creation project can also have the benefit of helping to refine research questions and it can be a vital aid to communication in larger-scale projects.

Good documentation is crucial to a data collection's long-term vitality: without it, the resource will not be suitable for future use and its provenance will be lost. Proper documentation contributes substantially to a data collection's scholarly value. At a minimum, documentation should provide information about a data collection's contents, provenance and structure, and the terms and conditions that apply to its use. It needs to be sufficiently detailed to allow the data creator to use the resource in the future, when the data creation process has started to fade from memory. It also needs to be comprehensive enough to enable others to explore the resource fully; and detailed enough to allow someone who has not been involved in the data creation process to understand the data collection and the process by which it was created. Apart from exceptional cases, the documentation that accompanies a data collection deposited with the History Data Service must meet the standards set out in their *Guidelines for Documenting Data* (available at <http://hds.essex.ac.uk/docguide.asp>).

Documenting any dataset is a resource-consuming task but GIS datasets are often more complicated. To start with, it is not always clear in a GIS what the documentation should refer to: should it be the whole dataset, each layer within the dataset, or even individual features within layers. The answer to this is often a combination of all three. While there is certainly a need for documentation about the dataset as a whole, individual layers or features may require further documentation to describe them.

The documentation required will vary between those captured from primary and secondary sources. It is impossible to produce definitive and exhaustive lists of the information required for both types of data, but primary source metadata is likely to need information on the following:

- the method of data capture
- the hardware and software used to capture the data (including version numbers)
- the resolution and accuracy of data capture
- an assessment of data quality
- a description of the purpose of the data capture
- who performed the data capture
- when the data were captured
- who owns the data.

For secondary sources, this list is complicated by the need to include information on both the original source and on the data capture. In this case the list might include:

- the purpose for which the data were captured
- the hardware and software used to capture the data (including version numbers)
- the purpose for which the source was produced
- the scale of the original source
- the method of data capture
- the scale or resolution at which the data were captured
- the root mean square (RMS) error of the data capture
- who owns the copyright on the original source
- the date when the data were captured or purchased
- who captured the data
- who owns the captured data.

Where additional work is done on an existing dataset, whether to derive a new dataset or merely update an existing one, this also needs to be documented. Here again the documentation would need to include:

- a description of what has been done to the data
- the hardware and software used to capture the data (including version numbers)
- the date of the work
- who did the work
- the relationship between this work and the dataset it was derived from.

9.5 FURTHER INFORMATION

Staff at the HDS will be happy to answer any queries you may have about creating, documenting and depositing data and can provide you with information about HDS training courses.

History Data Service, UK Data Archive, University of Essex, Colchester, CO4 3SQ. Tel: +44 (0)1206 873984. Fax: +44 (0)1206 872003. Email: hds@essex.ac.uk. URL: http://hds.essex.ac.uk

If your interest is in arts, humanities or social science data, rather than historical data *per se*, please contact either the appropriate service provider for the Arts and Humanities Data Service or the UK Data Archive. If you are in doubt regarding the appropriate AHDS service provider, please contact the AHDS Executive in the first instance.

AHDS Executive.
Tel: +44 (0)20 7928 7371.
Fax: +44 (0)20 7928 6825.
email: info@ahds.ac.uk.
URL: http://ahds.ac.uk/bkgd/exec.

Archaeology Data Service.
Tel: +44 (0)1904 433954.
Fax: +44 (0)1904 433939.
Email: info@ads.ahds.ac.uk.
URL: http://ads.ahds.ac.uk/

Oxford Text Archive.
Tel: +44 (0)1865 273238.
Fax: +44 (0)1865 273275.
Email: info@ota.ahds.ac.uk.
URL: http://ota.ahds.ac.uk/

Performing Arts Data Service.
Tel: +44 (0)141 330 4357/3810.
Fax: +44 (0)141 330 3659.
Email: info@pads.ahds.ac.uk.
URL: http://www.pads.ahds.ac.uk/

Visual Arts Data Service.
Tel: +44 (0)1252 892723.
Fax: +44 (0)1252 712925.
Email: info@vads.ahds.ac.uk.
URL: http://vads.ahds.ac.uk/

UK Data Archive.
Tel: +44 (0)1206 872001.
Fax: +44 (0)1206 872003.
Email: archive@essex.ac.uk.
URL: http://www.data-archive.ac.uk/

Section 10: Glossary and Bibliography

10.1 GLOSSARY

Accuracy: The difference between a set of representative values and the actual values. The accuracy of a *point* location would be the difference between the point's coordinates in the GIS and the coordinates accepted as existing in the real world. Section 3.4.

Ancillary documentation: Information that describes how the data were created or how they can be used. Section 9.1.

Animated GIF: A *GIF* is a bitmap file format often used on the World Wide Web. An animated GIF is a series of individual GIF frames joined together to create an *animation*. It is perhaps the easiest way to create and view simple animations. Section 6.7.

Animation: A collection of static images joined together and shown consecutively so that they appear to move. Section 6.7.

Arc: See *line*.

ArcInfo: Was the market leading GIS software package when GIS computing was workstation-based. Is now available for NT but has in some ways been superseded by desktop solutions such as its sister product *ArcView*, and *MapInfo*. *MapInfo* and *ArcView* are produced by Environmental Systems Research Institute (ESRI).

ArcView: A commonly used desktop GIS software package produced by Environmental Systems Research Institute (ESRI). Its sister product *ArcInfo* provides more functionality but is harder to use.

Area cartogram: These are *choropleth maps* that have been distorted so that the size of the *polygons* is not proportional to the polygon's area, but is instead proportional to another of the polygon's variables such as its total population. Section 6.6.

Areal interpolation: The process by which data from one set of source polygons are re-districted onto a set of overlapping but non-hierarchical target polygons. Section 4.7.

Areas: See polygons.

Attribute data: Data that relate to a specific, precisely defined *location*. The data are often statistical but may be text, images or multimedia. These are linked in the GIS to *spatial data* that define the location. Section 2.2 (see also sections 1.2 and 2.1).

Attribute querying: A *query* that extracts features from a layer based on the value of its attribute data: for example, 'select polygons with an unemployment rate greater than 15%' would be an attribute query. Section 4.2.

AVI: A video file format that can be used to publish *animations*. Section 6.7.

Blunder: The introduction of *error* by mistakes. Section 3.5.

Buffering: A buffer is a polygon that encloses all areas within a set distance of the spatial features. *Points*, *lines*, and *polygons* can all have buffers placed around them. For example,

Section 10: Glossary and Bibliography

if a user is interested in all areas within 1km of a church, a buffer would be placed around all the points representing churches. This would create a new layer consisting of polygons representing those areas within 1km of a church. Section 4.4.

Capture: See *data capture*.

Cartogram: See *area cartogram*.

Centroid: A point at the geometric centre of a *polygon*. This can be used to represent a *polygon* as a *point*. Section 2.3.

Choropleth maps: Maps of quantitative data that show patterns by using different colours or different shading for polygons classed in some way. For example, a map of *polygon*-based unemployment rates (expressed as percentages) might sub-divide rates into 0–5, 5–10, 10–15 and 15–20 and shade the polygons accordingly. Section 6.2.

Coordinate pair: An x and y coordinate used to represent a location in two-dimensional space, for example (6,4). Section 1.2.

Correlation: A form of statistical modelling that attempts to summarise how one dataset will vary in response to another. A correlation coefficient of +1.0 means that where there are high values in one set there will be high values in the other, while a correlation coefficient of –1.0 means that where there are high values in one set there will be low values in the other. A correlation coefficient of 0.0 means that there is no discernible relationship between the two sets. This is a form of *global analysis* as it only provides a single summary statistic for the entire study area. Section 7.2.

Coverage: See *layer*.

Dangling node: A *node* that should join with another node to join two or more *lines* together, but which does not join. This will result in holes in *topology*. Section 3.3.

Data capture: The process by which data are taken from the real-world (*primary source*), or from a secondary source such as a paper map, and entered into *GIS* software. From primary data this is usually through the use of *Global Positioning Systems* or *remote sensing*. For secondary data it is usually through *digitising* or *scanning*. Chapter 3 (and also section 1.4).

Database Management Systems: Software systems specifically designed to store *attribute data*. Section 1.2.

Date-stamping approach: A way of handling time in GIS where time is treated as an attribute. Each feature has date stamps attached that define the times that it was in existence. Section 5.4.

DBMS: See *Database Management Systems*.

DEM: See *Digital Terrain Model*.

DGPS: See *Differential GPS*.

Diachronic analysis: A form of analysis drawn from systems theory in which change over time is examined by comparing a large number of states, none of which are assumed to be in equilibrium. Section 5.2.

Differential GPS: A way of collecting *Global Positioning Systems* data with increased accuracy. It involves using a fixed base station at a known position to help find the location of a roving receiver. Section 3.8.

Digital Elevation Model: See *Digital Terrain Model*.

Digital Terrain Model: A data model that attempts to provide a three-dimensional representation of a continuous surface. Often used to represent relief. Section 2.5.

Digitising: In GIS this has a more precise meaning than in other disciplines. It usually refers to extracting coordinates from secondary sources such as maps to create vector data. Section 3.3.

Digitising table: A flat table with a fine mesh of wires under the surface used to allow accurate *digitising* of paper maps through the use of a *puck*. Section 3.3.

Digitising tablet: Similar to a *digitising table* only smaller. Section 3.3.

Dissolve: An operation in which adjacent *polygons* are merged if a selected feature of their *attribute data* are the same. An example might be merging polygons representing fields to create a new *layer* containing crop type. Section 4.4.

Drape: Involves laying features over a *digital terrain model* to provide information on features that lie on the terrain. The terrain model provides the shape of the terrain. Draped features may then include a *satellite image* of the terrain to show land-use, and *vector data* to show features such as roads. Section 2.5.

DTM: See *Digital Terrain Model*.

Ecological fallacy: The mistake of assuming that where relationships are found among aggregate data, these relationships will also be found among individuals or households. Section 7.2.

Edge-matching: See *rubber-sheeting*.

Error: In the context of *GIS* this means the difference between the real world and its digital representation. Section 3.5.

Error propagation: As layers of data are integrated through *overlays* the *error* present on the output *layer* will become the cumulative total of the *error* present on all the input layers. Section 4.6 (and also section 7.2).

Exploratory analysis: Statistical or visualisation techniques that attempt to produce a good summary of the data or the patterns with them. Section 7.2.

Fly-through: Often used to view *digital terrain models*. In a fly-through a user is given the functionality to allow him or her to move through the terrain in what appears to be three dimensions, thus giving the illusion of flying. It is an effective way of exploring a virtual landscape from different directions. Section 6.7.

Gazetteer: Often used to standardise place names or to locate place names within a hierarchy. These are often stored in a *Relational Database Management System*. Section 2.2.

GDA: See *Geographical Data Analysis*.

Geary's coefficient: A statistical technique that measures the degree of *spatial autocorrelation* present in the data. It is a form of *global analysis*. Section 7.3.

Geary's G_i: This is a *local analysis* form of *Geary's coefficient* that produces a measure of *spatial autocorrelation* for each *location* in the dataset. Section 7.3.

Geographical Data Analysis (GDA): A way of analysing data that explicitly incorporates information about *location* as well about attribute. This term may be used almost interchangeably with *spatial analysis*. Chapter 7.

Geographical Information Science: Methods of exploring and analysing spatially referenced data that take account of the benefits and limitations of such data. Section 1.2 (see also Chapter 7).

Geographical Information System: A computer system that combines *database management system* functionality with information about *location*. In this way it is able to capture, manage, integrate, manipulate, analyse and display data that are spatially referenced to the earth's surface. Chapter 1.

Section 10: Glossary and Bibliography

Geographically weighted regression (GWR): A form of *regression* modelling that explicitly incorporates the role of location. This is a form of *local analysis*. Section 7.3.

Geo-referencing: The process of proving a coordinate system to a *layer* of data. This often involves converting to a real-world coordinate system such as the British National Grid. Section 3.4.

GIF: Graphics Interchange Format. A bitmap graphics format from CompuServe which stores screen images economically and aims to maintain their correct colours even when transferred between different computers. GIF files are limited to 256 colours and like TIFFs, they use a lossless compression format but without requiring as much storage space.

GIS: See Geographical Information System.

GIS data: Data stored in a GIS are represented in two ways: *attribute data* says what the feature is, and *spatial data* says where it is using *points, lines, polygons*, or *pixels*. Section 1.2.

GISc: *Geographical Information Science*.

Global analysis: Forms of statistical analysis that provide an average measure of a relationship or relationships across the study area. Traditional *correlation* and *regression* techniques do this. They are flawed in that they do not allow for any geographical variations in the pattern so *local analysis* techniques are seen as more relevant in a GIS environment. Section 7.2.

Global Positioning Systems (GPS): A system based on satellites that allows a user with a receiver to determine precise coordinates for their location on the earth's surface. These are a *primary source* of *spatial data*. Section 3.8.

GPS: See *Global Positioning Systems*.

Graphic primitive: The basic representations of spatial features used in GIS. These are usually *points, lines, polygons* or *pixels*. Sections 1.4, 2.3 and 2.4.

GWR: See *Geographically weighted regression*.

Head-up digitising: The process by which *vector data* are extracted from *raster* scans using a cursor on-screen. Section 3.3.

Idrisi: A raster based GIS software package produced by Clark Labs, Clark University

Interpolation: A method of reallocating *attribute data* from one spatial representation to another. A simple example is to reallocate data from sample *points* to *polygons* using *Thiessen polygons*. *Kriging* is a more complex example that allocates data from sample points to a *surface*. Section 4.4.

Isolines: A *line* joining *points* of equal value. The most common example is the contour line on a map. Isobars showing lines of equal pressure on weather maps are another example. Section 6.2.

Java: A computer programming language often used to create Internet applications. Section 6.7.

JPEG: (Joint Photographic Experts Group), A digital image file format designed for maximal image compression. JPEG uses "lossy" compression in such a way that, when the image is decompressed, the human eye won't find the loss too obvious. The amount of compression is variable and the extent to which an image may be compressed without too much degradation depends partly on the image and partly on its use.

Key: In the context of *Relational Database Management Systems* this refers to a common field that can be used to join two or more tables. Section 2.2.

Key dates approach: A way of handling time in a GIS where the situation at different times is represented by different *layers*. Section 5.4.

Kriging: A form of statistical modelling that *interpolates* data from a known set of sample *points* to a continuous *surface*. Section 7.3.

Latitude: The angle of a location on the earth's surface from the equator expressed in degrees north or south. The Arctic Circle, for example, is at approximately latitude 66° north. Section 3.4.

Layer: The GIS data model represents the world by sub-dividing features on the earth's surface according to a specific theme. Each theme is then *georeferenced*. Examples of layers for a study area might include: roads, railways, urban areas, coal mines, etc. A layer usually consists of both *spatial* and *attribute* data. Section 2.6.

Line: A spatial feature that is given a precise location that can be described by a series of *coordinate pairs*. In theory a line has length but no width. Sections 1.2 and 2.3.

Local analysis: Forms of statistical analysis that allow relationships to vary across a study area by providing summary statistics for many locations. The results are usually best presented in map form. Examples of this type of technique include *Geary's Gi* and *Geographically Weighted Regression*. The opposite approach is *global analysis* where only a single summary statistic is provided for the average relationship across the study area. Section 7.2.

Location: The position of a feature on the earth's surface. In GIS this is usually explicitly defined in terms of precise coordinates. Chapters 1 and 2.

Location-allocation models: Models that attempt to find the optimum *location* for a feature based on information about other features. An example might be to find the best location for an industrial plant based on information about the transport network and the locations of raw materials and markets. Section 7.3.

Longitude: The angle of a location on the earth's surface usually expressed in degrees east or west of the Greenwich Meridian. New York, for example, is at approximately 74° west. Section 3.4.

Map algebra: A form of *overlay* used with *raster data*. In it the values for *pixels* on the output layer is calculated by performing a mathematical operation on the pixels from the input layers. The calculation may be arithmetic (addition, subtraction, multiplication, etc.) or Boolean (and, or, not, etc.). Section 4.6.

MapInfo: A commonly used desktop GIS software package produced by the MapInfo Corporation.

MAUP: See *Modifiable Areal Unit Problem*.

Metadata: Data that describe a dataset to allow others to find and evaluate it. Section 9.1.

Modifiable Areal Unit Problem (MAUP): Where data are published using totals for arbitrary areas such as administrative units, the patterns that they show may be simply the effect of the administrative units rather than genuine patterns among the underlying population. Section 7.2.

Moran's coefficient: A form of statistical modelling that measures the degree of *spatial autocorrelation* present in the data. Section 7.3.

MPEG: A video file format that can be used to publish *animations*. Section 6.7.

Network: A *topological* GIS data structure that uses a series of lines to describe, for example a transport or river network. Section 2.3 (see also section 7.3).

Network analysis: Usually used to analyse flows along a *network*. For example, to find the shortest path between two locations on a road network perhaps taking into account the different speeds and different fuel costs on different types of roads. Section 7.3.

Node: The start or end point of a *line* segment. As such a node is often the point at which lines intersect. Section 2.3.

Non-spatial data: See *attribute data*.

Object-orientated approach: A way of modelling the world that allocates entities to hierarchical classes. Section 2.1.

Overlay: A formal geometric intersection between two or more *layers* of spatially referenced data. A layer produced by an overlay will contain both the *spatial data* and the *attribute data* from the input layers. Section 4.6.

Pixels: The small units that sub-divide space to make up a *raster surface*. They are usually small grid squares. Sections 1.2 and 2.4.

Points: Spatial features that are given a precise *location* that can be described by a single *coordinate pair*. In theory a point has neither length nor width. Sections 1.2 and 2.3.

Polygons: Spatial features that are *areas* or *zones* enclosed by precisely defined boundaries. The boundaries of a polygon are formed from one or more *lines*. Sections 1.2 and 2.3.

Polyline: A term for a *line* used by some GIS packages.

Precision: The number of decimal places to which a value is given. This usually far exceeds its *accuracy*. For example, a GIS might give the coordinate of a *point location* for building to ten decimal places providing a value that is precise to fractions of a centimetre. In reality this value may only be accurate to the nearest ten meters. Section 3.5.

Primary source: In GIS terms this usually means a digital data source that is derived directly from the real world such as through *Global Positioning Systems* or *remote sensing*. Section 3.8.

Projection system: A method by which features on a curved earth are translated to be represented on a flat map sheet. This involves converting from *longitude* and *latitude* to x and y coordinates. Section 3.4.

Proximity measure: Usually an n by n matrix that gives a measure of the influence each *location i* has on each other *location j*. This is often expressed as a weighting W_{ij}. Section 7.3.

Puck: A hand held device used with a *digitising table* or *tablet*. It is used to point to an exact *location* in order to capture its coordinate. Section 3.3.

Quadrat analysis: Analysis where the study area is sub-divided into regular grid squares and the number of occurrences of a phenomenon in each square is counted. The resulting pattern can then be mapped. Quadrat analysis is not a particularly satisfactory technique as the results are too reliant on the size and position of the grid squares. Better techniques such as kernel estimations are described in the literature.

Quadtree: A way of encoding *raster data* that attempts to reduce storage requirements by avoiding sub-dividing homogeneous areas rather than storing values for every *pixel*. Section 2.4.

Quality: In the context of *GIS data*, quality usually refers to how fit the data are for a particular purpose. Section 3.5.

Querying: The process by which data are retrieved from a database in order to gain information from it. Section 4.2 (see also sections 2.2 and 2.3).

Raster data model: A way of representing the earth's surface by sub-dividing it into small *pixels*, usually square cells. Each pixel has values attached to it providing *attribute data* about the pixel. Section 2.4.

Raster-to-vector conversion: The process by which *vector* features (*points*, *lines* and *polygons*) are automatically extracted from *raster data*. This usually requires a large amount of user input and is often error prone. Section 3.7.

RDBMS: See *Relational Database Management Systems*.

Reference points: A small number of points used to *georeference* a *layer*. Often the four corners of the layer are used. Once the layer has been *digitised* we know the coordinates of the reference points in inches from the bottom left-hand corner of the *digitising table* or *tablet*. We also know their *locations* in real-world units from the map. This allows us to convert the entire layer's coordinates from digitiser inches to real-world coordinates. Section 3.4.

Regression: A form of statistical modelling that attempts to evaluate the relationship between one variable (termed the dependent variable) and one or more other variables (termed the independent variables). It is a form of *global analysis* as it only produces a single equation for the relationship thus not allowing any variation across the study area. *Geographically Weighted Regression* is a *local analysis* form of regression. Section 7.2.

Relational Database Management Systems: Software systems that store data in such a way that tables can be joined together by linking on a common item of data, termed a *key*. Section 2.2.

Relational join: The way by which two or more tables from a *Relational Database Management System* can be joined together based on one or more common items or *keys*. Section 2.2.

Remote sensing: The process by which *satellite images* are created by scanning the earth's surface using sensors on satellites. Section 3.8 (see also section 2.4).

RMS Error: See *Root Mean Square Error*.

Root Mean Square Error (RMS): A measure of the average *error* across a map. It is used in *digitising* to give an approximate measure of the difference between the real-world coordinates and the registration *points* on the digital *layer*. Section 3.5 (see also section 9.4).

Rubber-sheeting: The process by which a *layer* is distorted to allow it to be seamlessly joined to an adjacent layer. Often this has to be done when layers created from adjacent map sheets are joined together. It is a process that inevitably introduces some *error*. Section 4.3.

Run-length encoding: A way of encoding *raster data* that reduces storage requirements by creating linear groups of identical *pixels* rather than storing the values of each pixel individually. Section 2.4.

Satellite images: *Raster* models of the earth's surface produced from sensors on satellites. Section 3.8 (see also section 2.4).

Scanning: The process by which *raster* data is *captured* from paper maps. Section 3.2.

Segments: See *lines*.

Sliver polygons: Small *polygons* formed as a result of overlaying two or more *layers* of *vector data*. These are formed due to small differences in the way that identical *lines* have been *digitised*. Section 4.6.

Space: In a GIS context this means position on the earth's surface. Its meaning is very similar to *location*. Chapters 1 and 2.

Space-time composite: A way of handling time in GIS that preserves *topology* by sub-dividing space into a small set of areas that can then be re-aggregated into the arrangement that existed at different dates. Section 5.4.

Spans: A *raster* based GIS software package produced by PCI-Geomatics

Spatial analysis: A way of analysing data that explicitly incorporates information about *location* as well about *attribute*. This term may be used almost interchangeably with *geographical data analysis*. Chapter 7.

Spatial autocorrelation: The degree to which a set of features tend to be clustered together (positive spatial autocorrelation) or be evenly dispersed (negative spatial autocorrelation) over the earth's surface. This is often measured using either *Geary's coefficient* or *Moran's coefficient*. When data are spatially autocorrelated the assumption that they are independently random is invalid, so many statistical techniques are invalidated. Section 7.3.

Spatial data: Data that define a *location*. These are in the form of *graphic primitives* that are usually either *points, lines, polygons* or *pixels*. Sections 1.2, 2.3 and 2.4.

Spatial querying: A *query* that extracts features from a layer based on their location; for example, clicking on a *point* and listing its *attribute data* is a spatial query. Section 4.2.

SQL: See Structured Query Language.

Structured Query Language (SQL): A language used by many *Relational Database Management Systems* to manipulate their data. Section 2.2.

Surfaces: A surface is a way of modelling *space* that attempts to treat it as continuous rather than sub-dividing it into discrete features such as *polygons*. Surfaces are usually modelled either as *raster data* or *digital terrain models*. Sections 2.1, 2.4 and 2.5.

Synchronic analysis: A form of analysis drawn from systems theory in which change over time is examined by comparing the situation at two points in time when the system is assumed to be in equilibrium. Section 5.2.

Temporal data: Data that explicitly refer to time. Chapter 5.

Tessellation: A sub-division of *space* into discrete elements. *Raster surfaces* sub-divide space into regular tessellations such as *pixels*. *Polygons* are examples of irregular tessellations. Section 2.1.

Theme: See *layer*.

Thiessen polygons: A method of allocating *space* to the nearest *point*. The input *layer* will contain a set of points. The output layer, containing the Thiessen polygons, will contain *polygons* whose boundaries are *lines* of equal distance between two points. Section 4.4.

TIN: See *Triangular Irregular Network*.

Topology: The description of how spatial features are connected to each other. Section 2.3 (see also section 5.3 for the problems of creating topology that connects features through time as well as space).

Travelling Salesman Problem: A form of *network analysis* that attempts to find the shortest or cheapest route between a number of *locations* on a *network*. Section 7.3.

Triangular Irregular Network: A data structure that produces a continuous *surface* from *point* data. Often used to create a *digital terrain model*. Section 2.5.

Uncertainty: A measure of the amount of doubt or distrust with which the data should be used. Section 3.5.

Vector data model: Divides space into discrete features, usually *points, lines* or *polygons*. Section 2.3.

Vector-to-raster conversion: The process by which *vector data* are converted to *rasters*. This is usually automated. Section 3.7.

Voronoi diagrams: See *Thiessen polygons*.

Web-based mapping: Maps created for use on the Internet so they often have some interactive functionality. Web-based mapping is not well developed with *vector* file formats. Section 6.7.

Whole-map analysis: See *global analysis*.

Zones: See *polygons*.

10.2 BIBLIOGRAPHY

The AA: Travel Guides and Maps. [online]. Available from: http://www.theaa.com/services/bookshop_home.html [2 July 2002].

Anselin, L., 1999. Interactive techniques and exploratory spatial data analysis, in P.A. Longley, M.F. Goodchild, D.J. Maguire and D.W. Rhind (eds), *Geographical Information Systems: Principles, Techniques, Management and Applications*. 2nd edn. Chichester: John Wiley, 239–51.

Arts and Humanites Data Service Homepage. [online]. Available from: http://www.ahds.ac.uk/ [2 July 2002].

Association for Geographic Information (AGI). [online]. Available from: http://www.agi.org.uk/ [2 July 2002].

Australian Centre of the Asian Spatial Information and Analysis Network (ACASIAN). [online] Available from: http://www.asian.gu.edu.au/ [2 July 2002].

Bailey, T.C. and Gatrell, A.C., 1995. *Interactive Spatial Data Analysis*. Harlow: Longman.

Barnsley, M., 1999. Digital remotely-sensed data and their characteristics, in P.A. Longley, M.F. Goodchild, D.J. Maguire and D.W. Rhind (eds), *Geographical Information Systems: Principles, Techniques, Management and Applications*. 2nd edn. Chichester: John Wiley, 451–66.

Bartholomew: Mapping Solution Provider. [online]. Available from: http://www.bartholomewmaps.com/ [2 July 2002].

Bartley, K. and Campbell, B.M.S., 1997. Inquisitiones Post Mortem, GIS, and the creation of a land-use map of medieval England. *Transactions in GIS* 2, 333–46.

Bernhardsen, T., 1999. Choosing a GIS, in P.A. Longley, M.F. Goodchild, D.J. Maguire and D.W. Rhind (eds), *Geographical Information Systems: Principles, Techniques, Management and Applications*. 2nd edn. Chichester: John Wiley, 589–600.

Boonstra, O.W.A., 1994. Mapping the Netherlands, 1830 to 1994. The use of NLKAART. In: M. Goerke, ed. *Coordinates for Historical Maps*. Gottingen: Max-Planck-Institut fur Geschichte, 156–61.

Brunsdon, C., Fotheringham, A.S., and Charlton, M.E., 1996. Geographically Weighted Regression: A method for exploring spatial nonstationarity. *Geographical Analysis* 28, 281–98.

Bugayevskiy, L.M., and Snyder, J.P., 1995. *Map Projections: a Reference Manual*. London: Taylor & Francis.

Burrough, P.A. and McDonnell, R., 1998. *Principles of Geographical Information Systems for Land Resources Assessment*. 2nd edn. Oxford: Clarendon Press.

Calkins, H.W., 1990. Creating large digital files from mapped data, in: D.J. Peuquet and D.F. Marble (eds), *Introductory Readings in Geographic Information Systems*. London: Taylor & Francis, 209–14.

Camara, A.S. and Raper, J. (eds), 1999 . *Spatial Multimedia and Virtual Reality*. London: Taylor & Francis.

Campbell, J., 1987. *Introduction to Remote Sensing*. New York: Guildford Press.

Centre for Advanced Spatial Analysis. [online]. Available from: http://www.casa.ucl.ac.uk/ [2 July 2002].

The Charles Close Society for the study of Ordnance Survey maps. [online] Available from: http://www.charlesclosesociety.org.uk/ [2 July 2002].

China Historical GIS project. [online] Available from: http://www.people.fas.harvard.edu/~chgis/ [2 July 2002].

Chrisman, N.R., 1990. The accuracy of map overlays: a reassessment, in D.J. Peuquet and D.F. Marble, (eds), *Introductory Readings in Geographic Information Systems*. London: Taylor & Francis, 308–20.

Chrisman, N.R., 1990. Efficient digitizing through the combination of appropriate hardware and software for error detection and editing, in D.J. Peuquet and D.F. Marble (eds), *Introductory Readings in Geographic Information Systems*. London: Taylor & Francis, 233–46.

Chrisman, N.R., 1991. The error component in spatial data, in D.J. Maguire, M.F. Goodchild and D.W. Rhind (eds), *Geographical Information Systems: Principles and Applications. Volume 1: Principles*. Longman: Harlow, 1991, 165–74. Available from: http://www.wiley.co.uk/wileychi/gis/resources.html.

Chrisman, N.R., 1997. *Exploring Geographic Information Systems*. Chichester: Wiley.

Chrisman, N.R., 1999. What does 'GIS' mean? *Transactions in GIS* 3, 175–86.

Clarke, K.C., 1997. *Getting Started with Geographic Information Systems*. New Jersey: Prentice Hall.

Cliff, A.D. and Haggett, P., 1996. The impact of GIS on epidemiological mapping and modelling, in P. Longley and M. Batty (eds), *Spatial Analysis: Modelling in a GIS Environment*. GeoInformation International: Cambridge, 321–44.

Cliff, A.D. and Ord, J.K., 1981. *Spatial Processes: Models and Applications*. London: Pion.

Conclelis, H., 1999. Space, time, geography, in P.A. Longley, M.F. Goodchild, D.J. Maguire and D.W. Rhind (eds), *Geographical Information Systems: Principles, Techniques, Management and Applications*. 2nd edn. Chichester: John Wiley, 29–38.

Cowen, D.J., 1990. GIS versus CAD versus DBMS: what are the differences? in D.J. Peuquet and D.F. Marble (eds), *Introductory Readings in Geographic Information Systems*. London: Taylor & Francis, 52–62.

Curran, P., 1985. *Principles of Remote Sensing*. Harlow, Essex: Longman.

Dangermond, J., 1990. A review of digital data commonly available and some of the practical problems of entering them into a GIS, in: D.J. Peuquet and D.F. Marble, (eds), *Introductory Readings in Geographic Information Systems*. London: Taylor & Francis, 222–32.

Date, C.J., 1995. *An Introduction to Database Systems*. Reading, MA: Addison-Wesley.

Delano-Smith, C. and Kain, R.J.P., 1999, *English Maps: a History*. London: British Library.

Digital Chart of the World. [online] Available from: http://www.maproom.psu.edu/dcw/ [2 July 2002].

Dorling, D., 1992. Visualising people in space and time. *Environment and planning B*, 19, 613–37.

Dorling, D., 1993. Map design for census mapping. *Cartographic Journal* 30, 167–83.

Dorling, D., 1994. Cartograms for visualising human geography, in H.M. Hearnshaw and D.J. Unwin (eds), *Visualisation in GIS*. Chichester: John Wiley & Sons, 85–102.

Dorling, D., 1995. *A New Social Atlas of Britain*. London: John Wiley & Sons.

Dorling, D., 1996, *Area Cartograms: their Use and Creation*. Concepts and techniques in modern geography 59. Norwich: University of East Anglia, Environmental Publications.

Dorling, D. and Fairburn, D., 1997. *Mapping: Ways of Representing the World*. Harlow: Longman.

EDINA homepage. [online]. Available from: http://edina.ed.ac.uk/ [2 July 2002].

Egenhofer, M.J. and Golledge, R.G. (eds), 1998. *Spatial and Temporal Reasoning in Geographic Information Systems*. Oxford: Oxford University Press.

Egenhofer, M.J. and Herring, J.R., 1991. High-level spatial data structures for GIS, in D.J. Maguire,

M.F. Goodchild and D.W. Rhind (eds), *Geographical Information Systems: Principles and Applications. Volume 1: Principles*. Harlow: Longman, 1991, 227–37. Available from: http://www.wiley.co.uk/wileychi/gis/resources.html.

Electronic Cultural Atlas Initiative. [online]. Available from: http://www.ecai.org [2 July 2002].

Environmental Systems Research Institute, 1994a. *Arc/Info Data Management: Concepts, Data Models, Database Design, and Storage*. Redlands, CA.: Environmental Systems Research Institute Inc.

Environmental Systems Research Institute, 1994b. *Map Projections: Geo-referencing Spatial Data*. Redlands, CA: Environmental Systems Research Institute Inc.

Environmental Systems Research Institute, 1997. *Understanding GIS: the Arc/Info Method*. 4th edition. Cambridge: GeoInformation International.

ESRI Virtual Campus: GIS Education and Training on the Web. [online]. Available from: http://campus.esri.com/ [2 July 2002].

Evans, I.S., 1977. The selection of class intervals. *Transactions of the Institute of British Geographers* 2, 98–124.

Fischer, M.M., 1999. Spatial analysis: retrospect and prospect, in P.A. Longley, M.F. Goodchild, D.J. Maguire and D.W. Rhind (eds), *Geographical Information Systems: Principles, Techniques, Management and Applications*. 2nd edn. Chichester: John Wiley, 283–92.

Fischer, M.M., Scholten, H.J. and Unwin, D. (eds), 1996. *Spatial Analytical Perspectives on GIS*. London: Taylor & Francis.

Fisher, P.F., 1991. Spatial data sources and data problems, in D.J. Maguire, M.F. Goodchild and D.W. Rhind (eds), *Geographical Information Systems: Principles and Applications. Volume 1: Principles*. Longman: Harlow, 175–89. Available from: http://www.wiley.co.uk/wileychi/gis/resources.html.

Fisher, P.F., 1999. Models of uncertainty in spatial data, in P.A. Longley, M.F. Goodchild, D.J. Maguire and D.W. Rhind (eds), *Geographical Information Systems: Principles, Techniques, Management and Applications*. 2nd edn. Chichester: John Wiley, 191–205.

Flowerdew, R., 1991. Spatial data integration, in D.J. Maguire, M.F. Goodchild and D.W. Rhind (eds), *Geographical Information Systems: Principles and Applications. Volume 1: Principles*. Longman: Harlow, 375–87. Available from: http://www.wiley.co.uk/wileychi/gis/resources.html.

Flowerdew, R. and Green, M., 1994. Areal interpolation and types of data, in A.S. Fotheringham and P.A. Rogerson (eds), *Spatial Analysis and GIS*. London: Taylor & Francis, 121–45.

Fotheringham, A.S., 1997. Trends in quantitative methods I: stressing the local. *Progress in Human Geography* 21, 88–96.

Fotheringham, A.S. and Rogerson, P.A. (eds), 1994. *Spatial Analysis and GIS*. London: Taylor & Francis.

Fotheringham, A.S. and Wong, D., 1991. The modifiable areal unit problem in multi-variant statistical analysis. *Environment and Planning A*, 23, 1025–44.

Fotheringham, A.S., Brunsdon, C. and Charlton, M.E. 2000. *Quantitative Geography: Perspectives on Spatial Data Analysis*. London: Sage.

Galloway, J.A., 2000. Reconstructing London's distributive trade in the later middle ages: the role of computer-assisted mapping and analysis, in M. Woollard (ed.), *New Windows on London's Past: Information Technology and the Transformation of Metropolitan History*. Glasgow: Association for History and Computing, 1–24.

Gatley, D.A. and Ell, P.S., 2000. *Counting Heads: an Introduction to the Census, Poor Law Union Data and Vital Registration*. York: Statistics for Education.

Geographic Information System, Institute of Cultural Landscape Studies. [online] Available from: http://www.icls.harvard.edu/gis/contents.htm [2 July 2002].

Getis, A., 1999. Spatial statistics, in P.A. Longley, M.F. Goodchild, D.J. Maguire and D.W. Rhind (eds), *Geographical Information Systems: Principles, Techniques, Management and Applications*. 2nd edn. Chichester: John Wiley, 239–51.

GIS World Wide Web resource list. [online]. Available from: http://www.geo.ed.ac.uk/home/giswww.html [2 July 2002].
Gittings, B., 2002. *Digital Elevation Data Catalogue*. Available from: http://www.geo.ed.ac.uk/home/ded.html
Goodchild, M.F., 1987. *Introduction to spatial autocorrelation*. Concepts and Techniques in Modern Geography 47. Norwich: GeoAbstracts.
Goodchild, M.F., 1992. Geographical Information Science. *International Journal of Geographical Information Systems* 6, 31–45.
Goodchild, M.F. and Gopal, S., 1989, eds. *The Accuracy of Spatial Databases*. London: Taylor & Francis.
Goodchild, M.F. and Lam, N.S.-N., 1980 Areal interpolation: a variant of the traditional spatial problem. *Geo-processing* 1, 297–312.
Goodchild, M.F., Anselin, L. and Deichmann, U., 1993. A framework for the areal interpolation of socio-economic data. *Environment & Planning A*, 25, 383–97.
The Great Britain Historical GIS project. [online]. Available from: http://www.geog.port.ac.uk/gbhgis/ [2 July 2002].
Green, M. and Flowerdew, R., 1996, New evidence on the modifiable areal unit problem, in P. Longley and M. Batty (eds), *Spatial Analysis: Modelling in a GIS Environment*. Cambridge: GeoInformation International, 41–54.
Gregory, I.N., 2000. Longitudinal analysis of age- and gender-specific migration patterns in England and Wales: a GIS-based approach. *Social Science History* 24, 471–503.
Gregory, I.N., 2002. Time-variant GIS databases of changing historical administrative boundaries: A European comparison. *Transactions in GIS* 6 (2), 161–78.
Gregory, I.N., 2002. The accuracy of areal interpolation techniques: standardising 19th and 20th century census data to allow long-term comparisons. *Computers, Environment and Urban Systems* 26 (4), 293–314.
Gregory, I.N. and Southall, H.R., 1998. Putting the past in its place: The Great Britain Historical GIS, in S. Carver (ed), *Innovations in GIS 5*. London: Taylor & Francis, 210–21.
Gregory, I.N. and Southall, H.R., 2000. Spatial frameworks for historical censuses – the Great Britain Historical GIS, in P.K. Hall, R. McCaa and G. Thorvaldsen (eds), *Handbook of Historical Microdata for Population Research*. Minneapolis: Minnesota Population Center, 319–33.
Gregory, I.N., Southall, H.R. and Dorling, D., 2000. A century of poverty in England & Wales, 1898–1998: a geographical analysis, in J. Bradshaw and R.D. Sainsbury (eds), *Researching Poverty*. Aldershot: Ashgate, 130–59.
Guptil, S.C., 1999. Metadata and data catalogues, in P.A. Longley, M.F. Goodchild, D.J. Maguire and D.W. Rhind (eds), *Geographical Information Systems: Principles, Techniques, Management and Applications*. 2nd edn. Chichester: John Wiley, 677–92.
Haggett, P., 1979. *Geography: a Modern Synthesis*. 3rd ed. New York: Harper & Row.
Harder, C., 1998. *Serving Maps on the Internet: Geographic Information on the World Wide Web*. Redlands, CA: ESRI Press.
Harley, J.B., 1975, *Ordnance Survey Maps: a Descriptive Manual*. Southampton: Ordnance Survey.
Harris, T.M., 2000. Moving GIS: exploring movement in prehistoric cultural landscapes using GIS. In: G.R. Lock, ed. *Beyond the Map: Archaeology and Spatial Technologies*. Oxford: IOS Press, 116–23.
Harvey, C. and Press, J., 1996. *Databases in Historical Research: Theory, Methods and Applications*. London: Palgrave Macmillan.
Healey, R.G., 1991. Database management systems, in D.J. Maguire, M.F. Goodchild and D.W. Rhind (eds), *Geographical Information Systems: Principles and Applications. Volume 1: Principles*. Harlow: Longman, 251–67. Available from: http://www.wiley.co.uk/wileychi/gis/resources.html.

Healey, R.G. and Stamp, T.R., 2000. Historical GIS as a foundation for the analysis of regional economic growth: theoretical, methodological, and practical issues. *Social Science History* 24, 575–612.

Hearnshaw, H.M. and Unwin, D.J. (eds), 1994 *Visualization in GIS*. Chichester: John Wiley & Sons.

Heuvelink, G.B.M., 1999. Propagation of error in spatial modelling with GIS, in P.A. Longley, M.F. Goodchild, D.J. Maguire and D.W. Rhind (eds), *Geographical Information Systems: Principles, Techniques, Management and Applications*. 2nd edn. Chichester: John Wiley, 207–17.

Heywood, D.I., Cornelius, S. and Carver, S., 1998. *An Introduction to Geographic Information Systems*. Harlow: Longman.

Historical Atlas of Canada Online Learning Project. [online]. Available from: http://mercator.geog.utoronto.ca/hacddp/page1.htm [2 July 2002].

History Data Service. [online]. Available from: http://hds.essex.ac.uk [2 July 2002].

History Data Service, 2002, *Guidelines for Documenting Data*. [online]. Available from: http://hds.essex.ac.uk/docguide.asp [2 July 2002]

Hofmann-Wellenhof, B., Lichtenegger, H., and Collins, J., 1994. *Global Positioning Systems: Theory and Practice*. 4th edition. Wien: Springer.

Isaaks, E.H. and Srivastava, R.M., 1989. *An Introduction to Applied Geostatistics*. Oxford: Oxford University Press.

Jones, C. 1997. *Geographic Information Systems and Computer Cartography*. Harlow: Addison Wesley Longman.

Jones, K. and Duncan, C., 1996. People and places: the multilevel model as a general framework for the quantitative analysis of geographical data, in P. Longley and M. Batty (eds), *Spatial Analysis: Modelling in a GIS Environment*. Cambridge: GeoInformation International, 79–104.

Keates, J., 1989. *Cartographic Design and Production*. 2nd edition. Harlow: Longman Scientific & Technical.

Keene, D., 2000. Preface, in M. Woollard (ed), *New Windows on London's Past: Iinformation Technology and the Transformation of Metropolitan History*. Glasgow: Association for History and Computing, vii–ix.

Kennedy, M., 1996. *The Global Positioning System and GIS: an Introduction*. Chelsea, Mich.: Ann Arbor Press.

Kennedy, L., Ell, P.S., Crawford, E.M. and Clarkson, L.A., 1999. *Mapping the Great Irish Famine, an Atlas of the Famine Years*. Belfast: Four Courts Press.

KINDS: Knowledge Based Interfaces to National Data Sets. [online]. Available from: http://www.kinds.ac.uk/kinds [2 July 2002].

Knowles, A.K., 2000. Introduction. *Social Science History*, 24, 451–70.

Knowles, A.K., 2002, ed. *Past Time, Past Place: GIS for Historians*. ESRI Press.

Koussoulakou, A., 1994. Spatio-temporal analysis of urban air pollution, in A. MacEachren and F. Taylor (eds), *Visualisation in Modern Cartography*. Oxford: Elsevier Science, 243–67.

Koussoulakou, A. and Kraak, M., 1992. Spatio-temporal maps and cartographic communication. *Cartographic Journal* 29, 101–8.

Kraak, M-J., 1999. Visualising spatial distributions, in P.A. Longley, M.F. Goodchild, D.J. Maguire and D.W. Rhind (eds), *Geographical Information Systems: Principles, Techniques, Management and Applications*. 2nd edn. Chichester: John Wiley, 157–73.

Kristiansson, G., 2000. *Building a National Topographic Database*. [online]. Available from: http://www.geog.port.ac.uk/hist-bound/project_rep/NAD_more_info.htm [2 July 2002].

Lang, L., 1995. GIS supports Holocaust survivors video archive. *GIS World* 8(10), 42–5.

Langford, M., Maguire, D. and Unwin D.J., 1991. The areal interpolation problem: estimating population using remote sensing in a GIS framework, in I. Masser and M. Blakemore (eds), *Handling Geographical Information: Methodology and Potential Applications*. New York: Longman, 55–77.

Langran, G., 1989. A review of temporal database research and its use in GIS applications. *International Journal of Geographical Information Systems* 3, 215–32.

Langran, G., 1992. *Time in Geographic Information Systems*. London: Taylor & Francis.

Langton, J., 1972. Systems approach to change in human geography. *Progress in Geography* 4, 123–78.

Lee, J., 1996. Redistributing the population: GIS adds value to historical demography. *History and Computing* 8, 90–104.

Lillesand, T. and Kiefer, R., 1994. *Remote Sensing and Image Interpretation*. New York: John Wiley & Sons.

Longley, P. and Batty, M. (eds), 1996. *Spatial Analysis: Modelling in a GIS Environment*. Cambridge: GeoInformation International, 41–54.

Longley, P.A., Brooks, S.M., McDonnell, R., and MacMillan, B. (eds), 1998, *Geocomputation: a Primer*. Chichester: John Wiley & Sons.

Longley, P.A., Goodchild, M.F., Maguire, D.J. and Rhind, D.W. (eds), 1999. *Geographical Information Systems: Principles, Techniques, Management and Applications*. 2nd edition. Chichester: John Wiley.

Longley, P.A., Goodchild, M.F., Maguire, D.J. and Rhind, D.W., 2001. *Geographic Information Systems and Science*. Chichester: John Wiley.

Louis, R. (ed.), 1993. *The Historical Atlas of Canada, volume II: the land transformed, 1800 to 1891*. Toronto: University of Toronto Press.

MacDonald, B.M., and Black, F.A., 2000. Using GIS for spatial and temporal analysis in print culture studies: some opportunities and challenges. *Social Science History* 24, 505–36.

MacDonnell, R. and Kemp, K., 1995. *International GIS Dictionary*. Cambridge: GeoInformation International.

MacDougall, E., 1975. The accuracy of map overlays. *Landscape Planning* 2, 23–30.

MacEachren, A., 1994. Time as a cartographic variable, in H.M. Hearnshaw and D.J. Unwin (eds), *Visualisation in GIS*. Chichester: John Wiley & Sons, 115–30.

MacEachren, A.M., 1995. *How Maps Work: Representation, Visualisation and Design*. London: Guildford Press.

Maguire, D.J., 1991. An overview and definition of GIS, in D.J. Maguire, M.F. Goodchild and D.W. Rhind (eds), *Geographical Information Systems: Principles and Applications. Volume 1: Principles*. Harlow: Longman, 9–20. Available from: http://www.wiley.co.uk/wileychi/gis/resources.html.

Maguire, D.J., Goodchild, M.F. and Rhind, D.W. (eds), 1991. *Geographical Information Systems: Principles and Applications*. London: Longman Scientific and Technical. Available from: http://www.wiley.co.uk/wileychi/gis/resources.html.

Maling, D.H., 1991. Coordinate systems and map projection for GIS, in D.J. Maguire, M.F. Goodchild and D.W. Rhind (eds), *Geographical Information Systems: Principles and Applications. Volume 1: Principles*. Longman: Harlow, 135–46. Available from: http://www.wiley.co.uk/wileychi/gis/resources.html.

Martin, D., 1996a. *Geographic Information Systems and their Socio-economic Applications*. 2nd edition. Hampshire: Routledge.

Martin, D., 1996b. Depicting changing distributions through surface estimations, in P. Longley and M. Batty (eds), *Spatial Analysis: Modelling in a GIS Environment*. Cambridge: GeoInformation International, 105–22.

Massey, D., 1999. Space-time, 'science' and the relationship between physical geography and human geography. *Transactions of the Institute of British Geographers: new series* 24, 261–76.

Miller, D. and Modell, J., 1988. Teaching United States history with the Great American History Machine. *Historical Methods* 21, 121–34.

Miller, P. and Greenstein, D. 1997. *Discovering Online Resources Across the Humanities: a Practical Implementation of the Dublin Core*. Bath: UKOLN.

MIMAS: Manchester Information and Associated Services. [online]. Available from: http://www.mimas.ac.uk/ [2 July 2002].

Monmonier, M., 1996. *How to Lie With Maps*. Chicago: University of Chicago Press.

Monmonier, M.S., and Schnell, G.A., 1988. *Map Appreciation*. New Jersey: Prentice Hall.

Mooney, G., 2000. The epidemiological implications of reconstructing hospital catchment areas in Victorian London, in M. Woollard (ed.), *New Windows on London's Past: Information Technology and the Transformation of Metropolitan History*. Glasgow: Association for History and Computing, 47–74.

Odland, J., 1988. *Spatial Autocorrelation*. London: Sage.

Old Maps.co.uk homepage. [online]. Available from: http://www.old-maps.co.uk/ [2 July 2002].

Oliver, R.R., 1993, *Ordnance Survey Maps: a Concise Guide for Historians*. London: Charles Close Society.

Openshaw, S., 1984. *The Modifiable Areal Unit Problem*. Concepts and Techniques in Modern Geography 38. Norwich, Geobooks.

Openshaw, S., 1991. A view on the GIS crisis in geography or using GIS to put Humpty-Dumpty back together again. *Environment and Planning A*, 23, 621–28.

Openshaw, S., 2000, ed. *GeoComputation*. London: Taylor & Francis.

Openshaw, S. and Alvanides, S., 1999. Applying geocomputation to the analysis of spatial distributions, in P.A. Longley, M.F. Goodchild, D.J. Maguire and D.W. Rhind (eds), *Geographical Information Systems: Principles, Techniques, Management and Applications*. 2nd edn. Chichester: John Wiley, 267–82.

Openshaw, S. and Clarke, G., 1996. Developing spatial analysis functions relevant to GIS environments, in M.M. Fischer, H.J. Scholten and D. Unwin (eds), *Spatial Analytical Perspectives on GIS*. London: Taylor & Francis, 21–38.

Openshaw, S. and Rao, L., 1995 Algorithms for re-aggregating 1991 Census geography. *Environment and Planning A*, 27, 425–46.

Openshaw, S. and Taylor, P.J., 1979. A million or so correlation coefficients: three experiments on the modifiable areal unit problem, in N. Wrigley (ed.), *Statistical Applications in the Spatial Sciences*. London: Pion, 127–44.

Openshaw, S., Waugh, D., and Cross, A., 1994. Some ideas about the use of map animation as a spatial analysis tool, in H.M. Hearnshaw and D.J. Unwin (eds), *Visualisation in GIS*. Chichester: John Wiley & Sons, 131–38.

Ordnance Survey Maps – Britain's National Mapping Agency. [online]. Available from: http://www.ordsvy.gov.uk/ [2 July 2002]

Ott, T. and Swiaczny, F., 2002, in press. The analysis of cultural landscape change: a GIS approach for handling spatio-temporal data. *History and Computing*.

Owen, T. and Pilbeam, E., 1992. *Ordnance Survey: Map Makers to Britain since 1791*. London: HMSO.

Oxborrow, E.A., 1989. *Databases and Database Systems: Concepts and Issues*. 2nd edn. Bromley: Chartwell-Bratt.

Parkes, D. and Thrift, N., 1980. *Times, Spaces and Places: a Chronogeographic Perspective*. Chichester: John Wiley and Sons.

Pearson, A. and Collier, P., 1998. The integration and analysis of historical and environmental data using a Geographical Information System: landownership and agricultural productivity in Pembrokeshire c. 1850. *Agricultural History Review* 46, 162–76.

Perseus Digital Library. [online]. Available from: http://perseus.csad.ox.ac.uk/ [2 July 2002].

Peuquet, D.J., 1990. A conceptual framework and comparison of spatial data models, in D.J. Peuquet and D.F. Marble (eds), *Introductory Readings in Geographic Information Systems*. London: Taylor & Francis, 250–85.

Peuquet, D.J., 1994. It's about time: A conceptual framework for the representation of temporal dynamics in Geographic Information Systems. *Annals of the Association of American Geographers* 84, 441–61.

Peuquet, D.J., 1999. Time in GIS and geographical databases, in P.A. Longley, M.F. Goodchild, D.J. Maguire and D.W. Rhind (eds), *Geographical Information Systems: Principles, Techniques, Management and Applications*. 2nd edn. Chichester: John Wiley, 91–103.

Peuquet, D.J. and Boyle, A.R., 1990. Interactions between the cartographic document and the digitizing process, in D.J. Peuquet and D.F. Marble (eds), *Introductory Readings in Geographic Information Systems*. London: Taylor & Francis, 215–21.

Pickles, J.J. (ed.), 1995. *Ground Truth: the Social Implications of Geographic Information Systems*. New York: Guildford Press.

Pickles, J.J., 1999. Arguments, debates and dialogues: the GIS-social theory debate and the concern for alternatives, in P.A. Longley, M.F. Goodchild, D.J. Maguire and D.W. Rhind (eds), *Geographical Information Systems: Principles, Techniques, Management and Applications*. 2nd edn. Chichester: John Wiley, 49–60.

Pitternick, A., 1993. The historical atlas of Canada: the project behind the product. *Cartographica* 30, 21–31.

Public Record Office: The National Archives. [online]. Available from: http://www.pro.gov.uk/ [2 July 2002].

Raper, J., 2000. *Multidimensional Geographic Information Science*. London: Taylor & Francis.

Ray, B. 2001, Salem Witchcraft Accusations. [online]. Available from: http://jefferson.village.virginia.edu/%7Ebcr/salem/salem.html [2 July 2002].

Richardus, P., and Adler, R.K., 1972. *Map Projections for Geodesists, Cartographers and Geographers*. London: North Holland Publishing.

Robinson, W.S., 1950. Ecological correlations and the behaviour of individuals. *American Sociological Review* 15, 351–57.

Robinson, A.H., Morrison, J.L., Muehrcke, P.C., Kimerling, A.J. and Guptill, S.C., 1995. *Elements of Cartography*. 6th edn. Chichester: John Wiley & Sons.

Robson, B.T., 1969. *Urban Analysis*. Cambridge: Cambridge University Press.

Salem Witch Trials. [online]. Available from: http://www.iath.virginia.edu/salem/ [2 July 2002].

Shepherd, I.D.H., 1991. Information integration and GIS, in D.J. Maguire, M.F. Goodchild and D.W. Rhind, eds. *Geographical Information Systems: Principles and Applications. Volume 1: Principles*. Longman: Harlow, 337–60. Available from: http://www.wiley.co.uk/wileychi/gis/resources.html.

Shepherd, I.D.H., 1995. Putting time on the map: dynamic displays in data visualization, in P.F. Fisher, ed. *Innovations in GIS 2*. London: Taylor & Francis, 169–88.

Siebert, L., 2000. Using GIS to document, visualize, and interpret Tokyo's spatial history. *Social Science History* 24, 537–574.

Smith, D.M., Crane, G. and Rydberg-Cox, J., 2000. The Perseus Project: a digital library for the humanities. *Literary and Linguistic Computing* 15, 15–25.

Southall, H.R., Gregory, I.N. and Ell, P.S., 2001. *The Geography of Religion in 1851*. Available from: http://alpha4.iso.port.ac.uk:8001/geog/owa/gbd_atlas.page?tsect=LANG&tpage=REL_1851&poly_id=0 [2 July 2002].

Spence, C., 2000a. Computers, maps and metropolitan London in the 1690s, in M. Woollard (ed.), *New Windows on London's Past: Information Technology and the Transformation of Metropolitan History*. Glasgow: Association for History and Computing, 25–46.

Spence, C., 2000b. *London in the 1690s: a Social Atlas*. London: Centre for Metropolitan History, Institute of Historical Research.

Survivors of the Shoah Visual History Foundation. [online]. Available from: http://www.vhf.org/ [2 July 2002].

Taylor, P.J., 1990. Editorial comment: GIS. *Political Geography Quarterly* 9, 211–12.

Thrift, N., 1977. *An Introduction to Time Geography*. Concepts and techniques in modern geography 13. Norwich: Geo-Abstracts Ltd.

Tomlin, C.D., 1991. Cartographic modelling, in D.J. Maguire, M.F. Goodchild and D.W. Rhind, eds. *Geographical Information Systems: Principles and Applications. Volume 1: Principles*. Longman: Harlow, 361–74. Available from: http://www.wiley.co.uk/wileychi/gis/resources.html.

Townsend, S., Chappell, C. and Struijve, O., 1999. *Digitising History: a Guide to Creating Digital Resources from Historical Documents*. Oxford: Oxbow Books. Available from: http://hds.essex.ac.uk/g2gp/digitising_history/index.asp.

Tufte, E.R., 1990. *Envisioning Information*. Cheshire, Connecticut: Graphics Press.

UK Data Archive, 2002. Available from: http://www.data-archive.ac.uk/home/.

United States Geological Survey: EarthExplorer, 2002. Available from: http://edcsns17.cr.usgs.gov/EarthExplorer/.

Unwin, D.J., 1995. Geographic Information Systems and the problem of 'error and uncertainty'. *Progress in Human Geography* 19, 549–58.

Unwin, D., 1996. Integration through overlay analysis, in M. Fischer, H.J. Scholten and D. Unwin (eds), *Spatial Analytical Perspectives on GIS*. London: Taylor & Francis, 129–38.

The Valley of the Shadow contents page, 2002. Available from: http://jefferson.village.virginia.edu/vshadow2/contents.html.

Vanhaute, E., 1994. The quantitative database of Belgian municipalities (19th/20th centuries): From diachronic worksheets to historical maps, in M. Goerke (ed.), *Coordinates for Historical Maps*. Gottingen: Max-Planck-Institut fur Geschichte, 162–75.

Veregin, H., 1999. Data quality parameters, in P.A. Longley, M.F. Goodchild, D.J. Maguire and D.W. Rhind (eds), *Geographical Information Systems: Principles, Techniques, Management and Applications*. 2nd edn. Chichester: John Wiley, 177–89.

Wachowicz, M., 1999. *Object-Orientated Design for Temporal GIS*. London: Taylor & Francis.

Weibel, R., and Heller, M., 1991. Digital terrain modelling, in D.J. Maguire, M.F. Goodchild and D.W. Rhind, eds. *Geographical Information Systems: Principles and Applications. Volume 1: Principles*. Harlow: Longman, 269–97.

Weibel, S., Godby, J., Miller, E. and Daniel, R. 1995. *OCLC/NCSA Metadata Workshop Report*. Dublin: Online Computer Library Centre.

Wolf, P.R., 2000. *Elements of Photogrammetry: with Applications in GIS*. 3rd edn. London: McGraw-Hill.

Woods, R. and Shelton, N., 1997. *An Atlas of Victorian Mortality*. Liverpool: Liverpool University Press.

Worboys, M.F., 1995. *GIS: A computing perspective*. London: Taylor & Francis.

Worboys, M.F., 1999. Relational databases and beyond, in P.A. Longley, M.F. Goodchild, D.J. Maguire and D.W. Rhind (eds), *Geographical Information Systems: Principles, Techniques, Management and Applications*. 2nd edn. Chichester: John Wiley, 373–84.

Wrigley, N., Holt, T., Steel, D. and Tranmer, M. 1996. Analysing, modelling, and resolving the ecological fallacy, in P. Longley and M. Batty (eds), *Spatial Analysis: Modelling in a GIS Environment*. Cambridge: GeoInformation International, 25–40.